ORBITAL SYMMETRY

A Problem-Solving Approach

ORBITAL SYMMETRY

A PROBLEM-SOLVING APPROACH

Roland E. Lehr and Alan P. Marchand
The University of Oklahoma

ACADEMIC PRESS *New York and London*

ACADEMIC PRESS, INC.
111 Fifth Avenue, New York, New York 10003

United Kingdom Edition published by
ACADEMIC PRESS, INC. (LONDON) LTD.
24/28 Oval Road, London NW1 7DD

Library of Congress Catalog Card Number: 72-159625

PRINTED IN THE UNITED STATES OF AMERICA

To Karen,嘉琛, Sara, and Boo Boo

CONTENTS

PREFACE

Rarely has a development in organic chemistry been greeted with the enthusiasm accorded Woodward and Hoffmann's approach to pericyclic reactions. The response may be gauged in terms of the development of alternative theoretical methods and the vast number of publications describing experimental results in the area, their content attesting to the wide applicability of the approach.

The definitive theoretical treatise by Woodward and Hoffmann appeared in 1970. Our book was born of the necessity of presenting this theory to our students. Hence the tone is introductory, and the book is addressed primarily to an audience of advanced undergraduate and beginning graduate students.

We have sought to familiarize the reader with several of the more often encountered methods of analyzing pericyclic reactions, and these methods should enable him to analyze virtually all of them. Problem solving is the foundation of our approach. Both the introductory and theory sections include problems to prepare the reader for the more extensive chapters of problems that follow. All problems (except those in Chapter VIII) are answered in the text and are fully referenced where appropriate. Many of the problems require the use of molecular models if they are to be appreciated. Prentice-Hall's "Framework Molecular Models" and Benjamin's "Maruzen Models" are best suited for the construction of the highly strained molecules often encountered in the problems, and we recommend their use.

Literature continues to accrue in the area of pericyclic reactions at such a rapid pace that we have been forced to select particular examples from the vast array available. Our judgment in selecting examples was strongly influenced by our desire to emphasize that literature which was current at the time of the preparation of our manuscript. Inevitably, much older work and even some important current examples could not be included. Such exclusions reflect restriction of space rather than negative judgment on our part.

We extend thanks to our numerous colleagues and students who have assisted in the preparation of this manuscript.

ROLAND E. LEHR
ALAN P. MARCHAND

Part I

INTRODUCTION AND THEORY

Chapter I

INTRODUCTION

Although the usefulness of molecular orbital theory as a device for analyzing a wide variety of chemical and spectroscopic behavior has been recognized for some time, it was not until 1965 that general awareness of its applicability to concerted reactions was awakened. Concerted reactions are those in which a reactant is converted to a product without intervention of an intermediate. Bond formation and bond breakage occur synchronously, though not necessarily symmetrically.

In 1965, Woodward and Hoffmann initiated a series of publications that has aroused the interest of a considerable body of chemists. They showed that the striking chemical behavior of three special classes of concerted reactions could be understood through application of molecular orbital theory. In particular, they were able to explain the fact that certain conjugated polyenes suffered thermal ring closure in one stereochemical sense, whereas the related conjugated polyene with either one more or one less double bond suffered thermal ring closure stereospecifically in an *exactly opposite* stereochemical sense. Furthermore, their analysis rationalized the intriguing consequences of changing the "reagent" for these transformations from heat to light, namely, that for a given conjugated polyene photochemical transformation led to one stereochemical result and thermal transformation led to the precise opposite! Similar differences in reactions involving intramolecular bond migrations along conjugated systems and in reactions involving intermolecular interactions of pi-systems to form cyclic structures were explained. The three general reaction types are cited in Fig. I.1, with one example of each.

In this chapter, we shall define each of the three types of reactions and discuss the stereochemical features of each. We shall present rules that have been derived which correlate their behavior and apply them to predict transformations within each set of reactions. Finally, in Section D, we shall find that electrocyclic, sigmatropic, and cycloaddition reactions are all representatives of a larger class of concerted reactions termed *pericyclic reactions*, and we shall discuss them in that context.

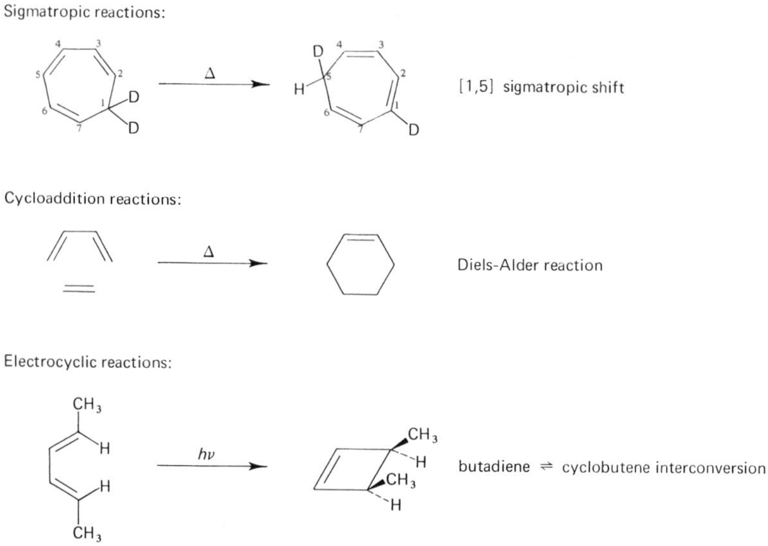

Sigmatropic reactions:

[1,5] sigmatropic shift

Cycloaddition reactions:

Diels-Alder reaction

Electrocyclic reactions:

butadiene ⇌ cyclobutene interconversion

Fig. I.1. *Sigmatropic, cycloaddition, and electrocyclic reactions.*

Some reactions, termed *symmetry forbidden*, confront very large energy barriers to reaction, whereas others, termed *symmetry allowed*, proceed with relative ease. In Chapter II, we shall explore the theoretical basis which differentiates these two classes of reactions, and thereby lay the foundation for problems which the reader will encounter in succeeding chapters.

A. Electrocyclic Reactions

Electrocyclic reactions were the first to be treated by Woodward and Hoffmann in their classic series of articles. The reactions are defined as involving the cyclization of an n pi-electron system to an $(n - 2)$ pi- + 2 sigma-electron system or the reverse process (Fig. I.2). Usually, the reactions are reversible and the

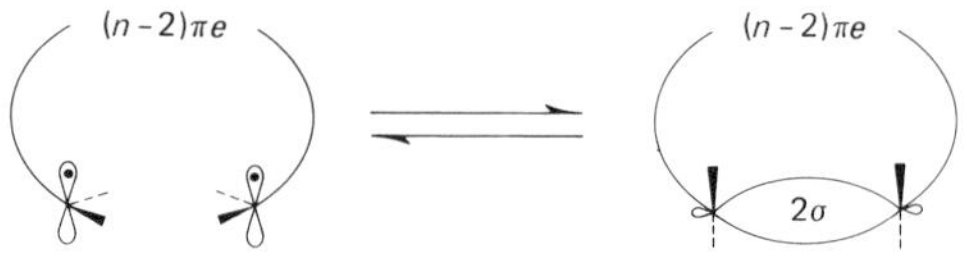

Fig. I.2. *Generalized representation of an electrocyclic reaction.*

observance of ring opening or ring closure will depend upon the thermodynamic stability of the open and closed forms. An example of an electrocyclic process is the conversion of 1,3,5-hexatriene to 1,3-cyclohexadiene [Eq. (I.1)].

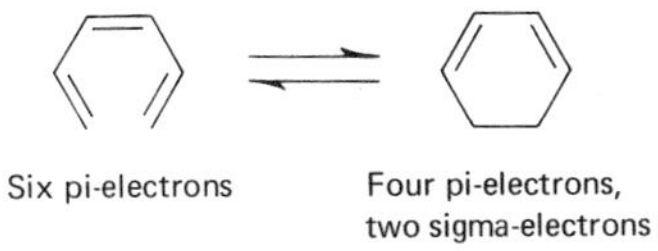

Six pi-electrons ⇌ Four pi-electrons, two sigma-electrons (I.1)

Woodward and Hoffmann pointed out that two distinct alternative cyclization (or ring-opening) modes are possible in these systems. They can be distinguished when the terminal positions are appropriately substituted, since as the $2p$ orbitals on the terminal carbon atoms rotate and form sigma-bonds, the groups attached to those carbon atoms must rotate, also (Fig. I.3). They are termed *disrotatory*

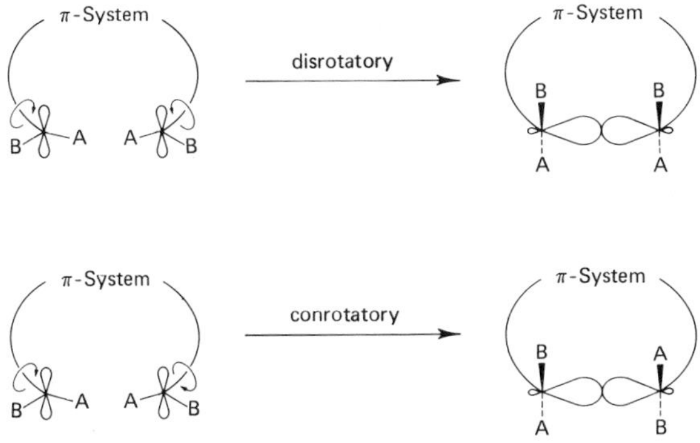

π-System disrotatory π-System

π-System conrotatory π-System

Fig. I.3. *Conrotatory and disrotatory motions.*

(pronounced dis′-rō-tā′-tôr-i) and *conrotatory* (pronounced kon′-rō-tā′-tôr-i), depending upon whether the terminal $2p$ pi-orbitals are rotated in the opposite or the same sense, as shown in the figure. The reader will note that, for the particular polyene depicted in Fig. I.3, disrotatory closure leads to a *cis* ring fusion whereas conrotatory closure leads to a *trans* ring fusion. Some examples of electrocyclic reactions are cited in [Eqs. (I.2), (I.3), and (I.4)].

Rules that have been established for electrocyclic reactions are listed in Table I.1. The reader should note that the systems are separated into two groups, those containing $4q$ and those containing $(4q + 2)$ electrons. The preferred stereo-

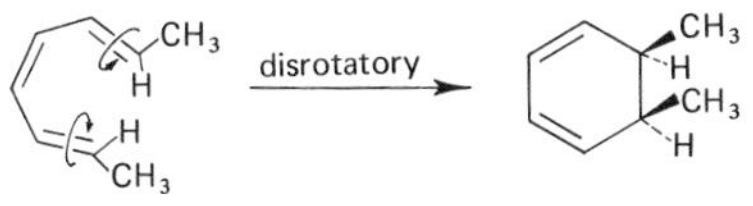

disrotatory (I.2)

disrotatory (I.3)

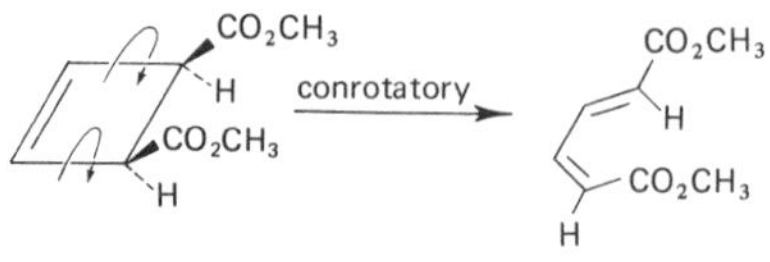

conrotatory (I.4)

Table I.1 *Selection Rules for Electrocyclic Reactions*[a]

n	Thermally allowed, photochemically forbidden	Thermally forbidden, photochemically allowed
$4q$	Conrotatory	Disrotatory
$4q + 2$	Disrotatory	Conrotatory

[a] q = integer.

chemical mode of transformation within each group depends upon whether light or heat is used to effect the reaction. Furthermore, the precisely opposite behavior of the two groups toward a given reagent (heat or light) can be noted. We shall find later (Chapter II) that this curious behavior can be explained by analyzing a particular feature of the orbitals involved in the reactions. As an example of the application of the rules in Table I.1, consider the *thermal* conversion of cyclobutene to butadiene: since the process involves four electrons ($n = 4 = 4q$, where $q = 1$) and is thermally induced, it is predicted to occur in the conrotatory manner.

PROBLEM I.1

Have the following reactions proceeded in the conrotatory or disrotatory manner?

(a)

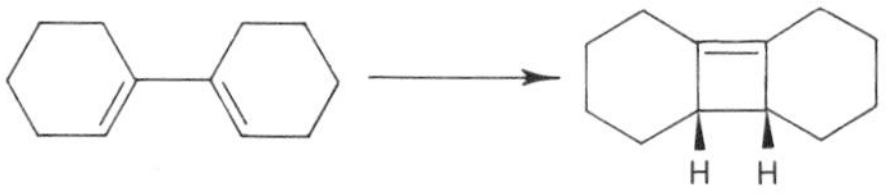

(b)

(c)

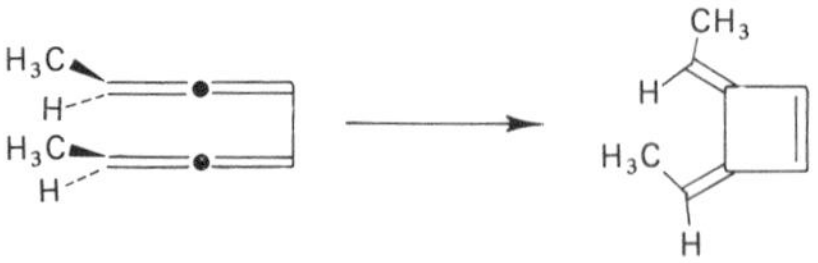

PROBLEM I.2

In some cases, more than one conrotatory or disrotatory pathway is available. Show that the cyclobutene below can open by two alternative conrotatory processes. What is the product in each instance? Do you expect them to be formed in equal amounts?

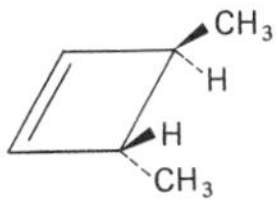

PROBLEM I.3

Review the reactions in Eqs. (I.2), (I.3), and (I.4) and Problem I.1. Determine whether the reactions shown, if concerted, should proceed under thermal or photochemical influence.

B. Sigmatropic Reactions

These reactions are defined as involving migration of a sigma-bond that is flanked by one or more conjugated systems to a new position within the system. The system is numbered by starting at the atoms attached to the migrating bond [Eq. (I.5)].

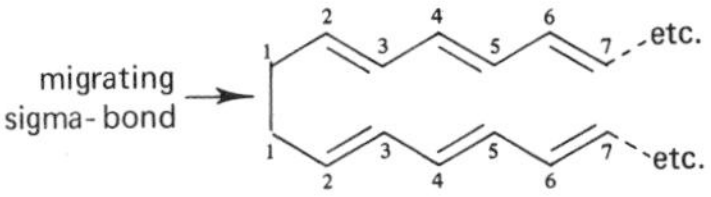

migrating sigma-bond (I.5)

The reaction is termed an [*i,j*] sigmatropic shift when the bond migrates from position [1,1] to position [*i,j*]. For example, the following reactions may be classified as shown [Eqs. (I.6) and (I.7)]:

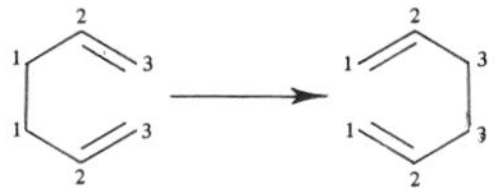

[3,3] sigmatropic shift (I.6)

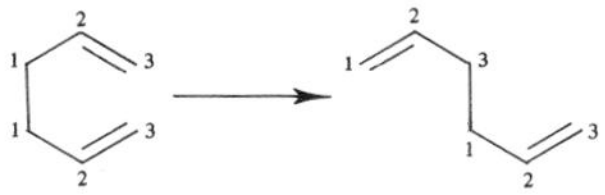

[1,3] sigmatropic shift (I.7)

PROBLEM I.4

Classify the following as sigmatropic reactions of order [*i,j*].

(a)

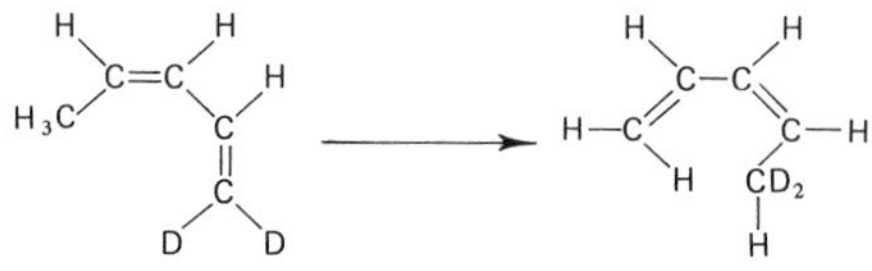

(b)

(c)

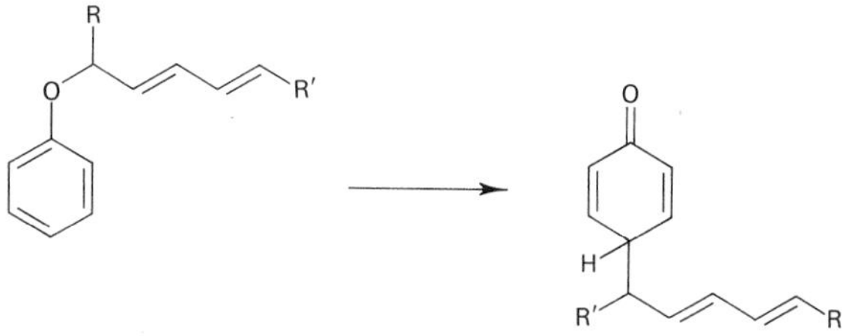

In sigmatropic reactions, the sigma-bond migrates across a conjugated network of atoms. In principle, the new bond could be formed on the same or on the opposite face of the pi-framework. These processes are termed *suprafacial* and *antarafacial*, respectively, and are illustrated for hydrogen migrations in Fig. I.4. Clearly, suprafacial migrations will usually be more feasible geometrically than antarafacial migrations; nevertheless, when the length of the conjugated chain becomes large enough, it becomes possible for a bond to migrate from the top face to the bottom face of the conjugated system.

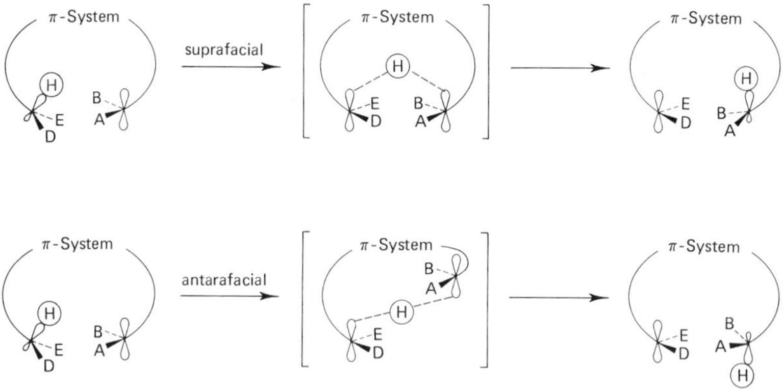

Fig. I.4. *Suprafacial and antarafacial hydrogen migrations.*

PROBLEM I.5

Classify the order of sigmatropic shift that produced each of the following products, (A) and (B). Indicate for each product whether it is derived from antarafacial or suprafacial migration.

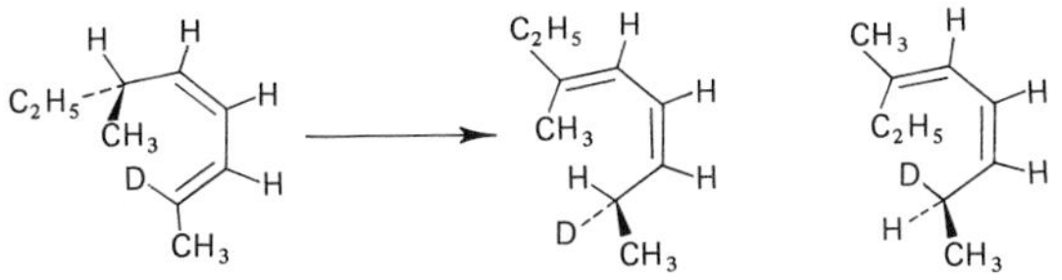

(A) (B)

Of further interest is the possibility for inversion at the migrating center. The reader will recall that inversion occurs at a carbon atom when a new bond is formed at the backside of the bond broken in the reaction, as can be seen in the S_N2 reaction of (*S*)-(+)-2-bromobutane with iodide ion in Fig. I.5. Of course, the option of inversion is not available to a hydrogen atom, and it can be experimentally verified for carbon only when the carbon atom is asymmetric. The ability of carbon to invert leads to four possible migration routes: suprafacial with retention, suprafacial with inversion, antarafacial with retention, and antarafacial with inversion (Fig. I.6). The reader will note that the processes are distinct, and that the products are different in each instance. Though these are the *only types* of migration possible, there is an additional route for each type which involves utilization of the other face of the pi-system.

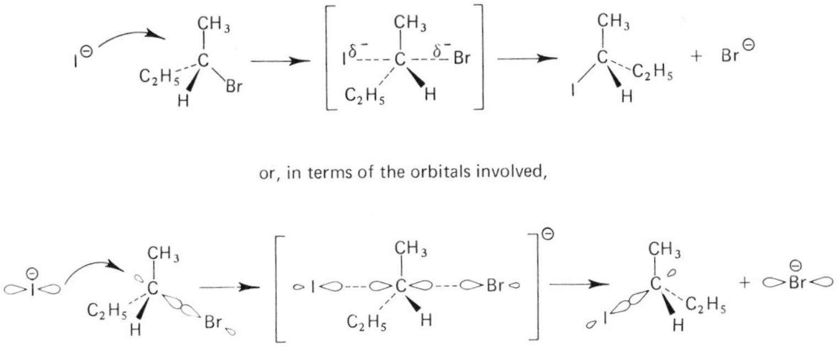

or, in terms of the orbitals involved,

Fig. I.5. *Inversion at carbon in an S_N2 reaction.*

PROBLEM I.6

Demonstrate that there is another "suprafacial with retention" and another "antarafacial with inversion" migration route available to the system in Fig. I.6 and that the products obtained are different from those obtained in the corresponding processes shown in Fig. I.6.

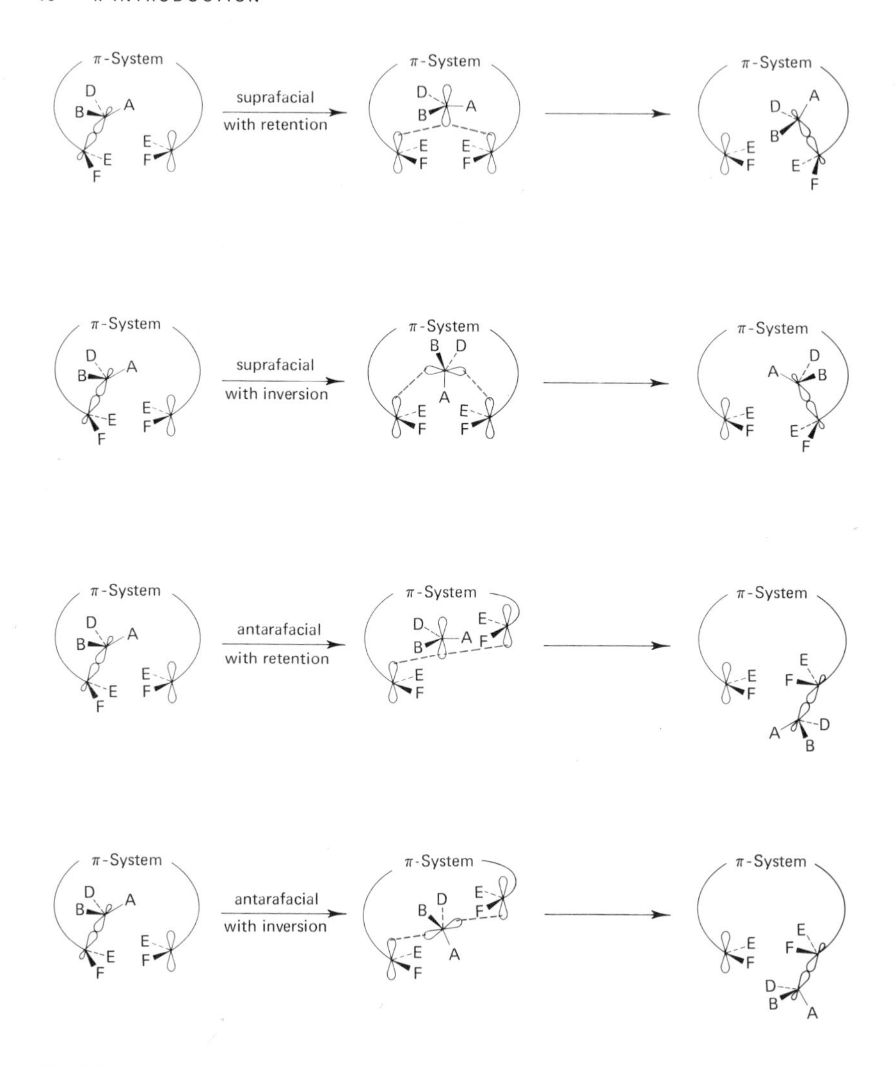

Fig. I.6. *The four possible types of sigmatropic migration modes for carbon.*

Despite the severe crowding that must be present in the transition state, inversion at the migrating center has been observed for suprafacial migrations and should be considered as a possibility. For hydrogen migrations, the rules given in Table I.2 have been developed. Thus, the well-known suprafacial [1,5] sigmatropic hydrogen shift is allowed thermally $[(1 + j) = 6 = (4n + 2)]$, as is observed (see Problem I.5). However, the suprafacial [1,3] hydrogen shift is forbidden thermally, but is allowed photochemically $[(1 + j) = 4 = 4n]$. This prediction is also in accord with experimental observation.

Table I.2 *Rules for Hydrogen Migrations of Order* $[i,j]^a$

$1 + j$	Suprafacial	Antarafacial
$4n$	Thermally forbidden, photochemically allowed	Thermally allowed, photochemically forbidden
$4n + 2$	Thermally allowed, photochemically forbidden	Thermally forbidden, photochemically allowed

[a]n = integer.

PROBLEM I.7

Classify the following reaction as a sigmatropic change of order $[i, j]$ and indicate whether the thermally induced reaction should be allowed suprafacially or antarafacially.

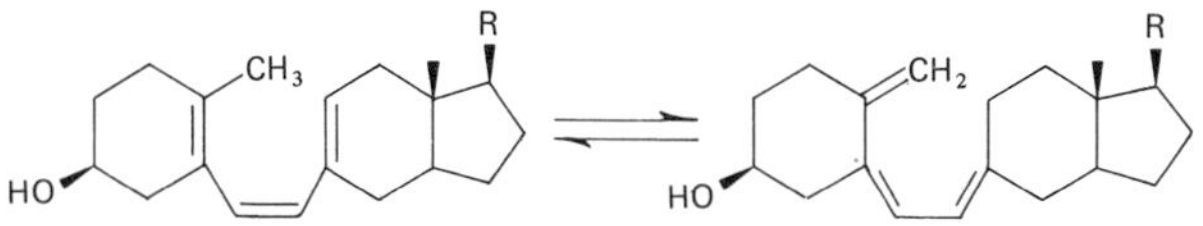

If the migrating hydrogen atom is replaced by a carbon atom, $[1, j]$ sigmatropic shifts will still obey the above rules (Table I.2), *provided that the group migrates with retention.* However, if the group migrates with inversion, the rules are precisely *reversed*. Thus, a suprafacial $[1,3]$ shift is allowed thermally, *if the migrating group suffers inversion.* For $[i, j]$ sigmatropic shifts in which both $[i, j] > 1$, it is necessary to specify whether the migration is suprafacial or antarafacial for *each* conjugated system involved. The rules derived for them are found in Table I.3.

PROBLEM I.8

Classify the following reaction as an $[i, j]$ sigmatropic shift and indicate whether the transformation has occurred suprafacially or antarafacially on each pi-component. Should the reaction be photochemically or thermally allowed?

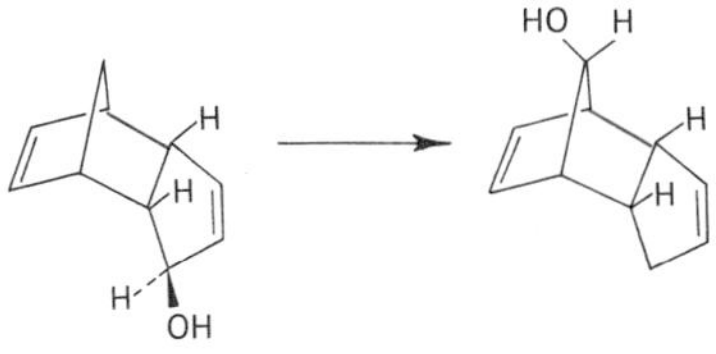

PROBLEM I.9

In the following *photochemical* conversion, designate the expected position of deuterium in the product (assume retention at the migrating group). Do you expect deuterium to be found at the α or γ position, or both?

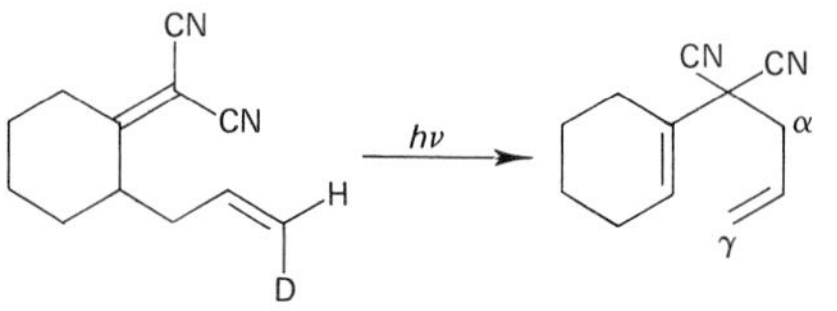

CN CN H D $h\nu$ CN CN α γ

Sigmatropic shifts within charged species are predicted to conform to a set of rules, as well. For $[1,j]$ sigmatropic shifts in which the migrating group shifts with retention, the rules are presented in Table I.4. Suprafacial $[1,2]$ sigmatropic shifts in carbonium ions are very well known $[(1 + j) = 3 = (4n + 3)]$. For reactions in which the migrating group undergoes inversion, the rules are precisely reversed.

PROBLEM I.10

In the following reactions, should the migrating carbon atom shift with retention or inversion?

(a)

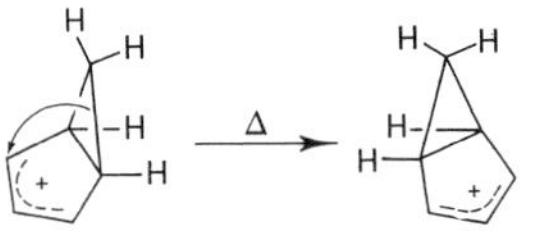

(b)

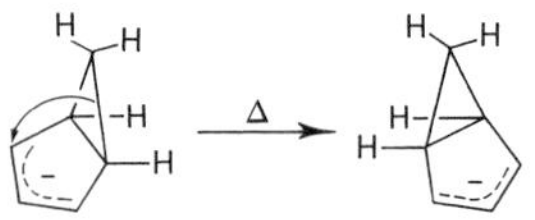

PROBLEM I.11

On the basis of your answers to parts (a) and (b) in Problem I.10, predict product stereochemistries for:

(a)

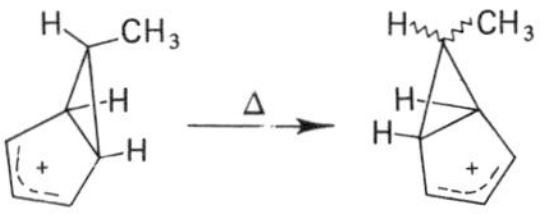

(b)

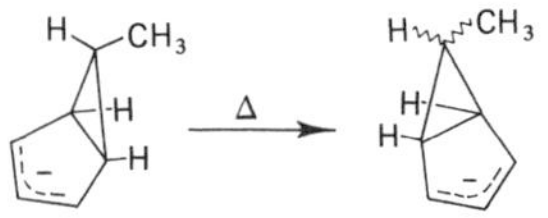

Though the rules for sigmatropic reactions appear complex at this time, we shall find in Section D that they can be summarized in one simple rule once we examine the involvement of the sigma-bond in more detail. It is worth noting here, though, that the dichotomy of behavior with respect to the influence of heat and light we noted for electrocyclic reactions persists in sigmatropic reactions. Furthermore, an examination of Tables I.3 and I.4 reveals that the stereo-

Table I.3 *Rules for* $[i,j]$ *Sigmatropic Shifts* $(i,j > 1)$

$i + j$	Antara–antara or supra–supra	Antara–supra or supra–antara
$4n$	Thermally forbidden, photochemically allowed	Thermally allowed, photochemically forbidden
$4n + 2$	Thermally allowed, photochemically forbidden	Thermally forbidden, photochemically allowed

chemistry of the preferred mode of migration is determined by the *number of electrons* involved in the process. For example, if we consider $[1,j]$ sigmatropic shifts, the number of electrons involved in a [1,5] shift in a neutral system [a $(1 + j) = 6 = (4n + 2)$ system] is 6 (2 sigma + 4 pi), which is equal to the number of electrons involved in a [1,6] shift in a cation [a $(1 + j) = 7 = (4n + 3)$ system]. The tables reveal that the stereochemical behavior of a neutral $(4n + 2)$ system should be the same as a *cationic* $(4n + 3)$ system.

Table I.4 *Rules for* $[1,j]$ *Sigmatropic Shifts for Charged Species*[a]

$1 + j$	Mode of migration	Cation	Anion
$4n + 1$	Suprafacial	Thermally forbidden, photochemically allowed	Thermally allowed, photochemically forbidden
	Antarafacial	Thermally allowed, photochemically forbidden	Thermally forbidden, photochemically allowed
$4n + 3$	Suprafacial	Thermally allowed, photochemically forbidden	Thermally forbidden, photochemically allowed
	Antarafacial	Thermally forbidden, photochemically allowed	Thermally allowed, photochemically forbidden

[a] n = integer.

C. Cycloaddition Reactions

The most easily visualized cycloaddition reactions are those that involve the addition of a system of m pi-electrons to a system of n pi-electrons to produce a new ring. Such reactions are termed $[m + n]$ cycloaddition reactions, of which the prototype is the Diels-Alder reaction [Eq. (I.8)].

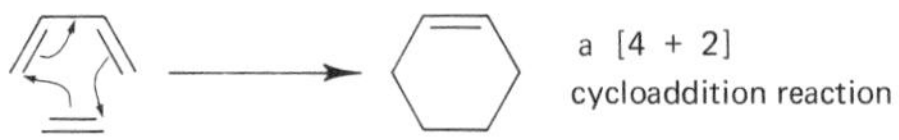

a [4 + 2] cycloaddition reaction (I.8)

The butadiene moiety contains four pi-electrons and the ethylene moiety contains two. The reaction is therefore designated a $[4 + 2]$ cycloaddition reaction. If more than two components are involved in the reaction, it is termed an $[m + n + \cdots]$ cycloaddition reaction, depending on the number of components involved. For example, in the reaction of norbornadiene with ethylene [Eq. (I.9)], there are three pi-components involved.

(I.9)

Two of them are held rigidly in the norbornadiene molecule, and the other is the ethylene double bond. Each component contains two pi-electrons, so the reaction is termed a $[2 + 2 + 2]$ cycloaddition reaction.

PROBLEM I.12

Classify the following as $[m + n + \cdots]$ cycloaddition reactions:

(a) CO O CO → CO O CO

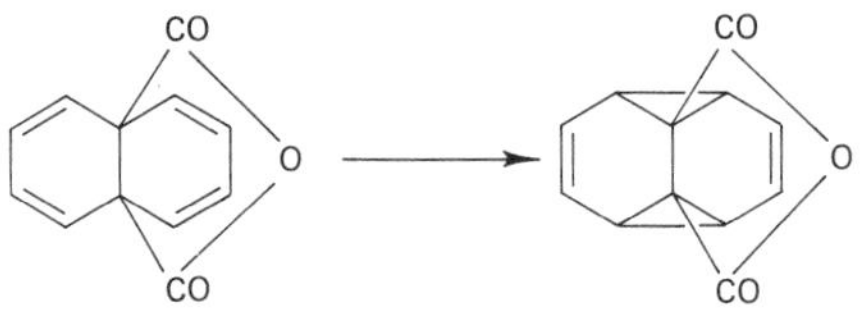

(b) +

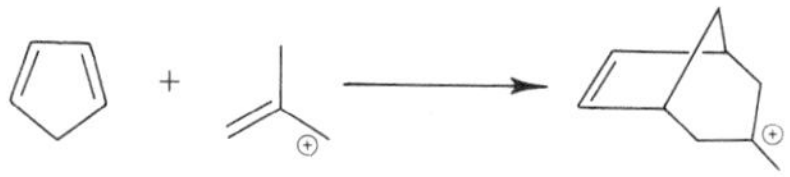

(c) R O O H R₁ H R₂ → R O O + H R₂ R₁ H

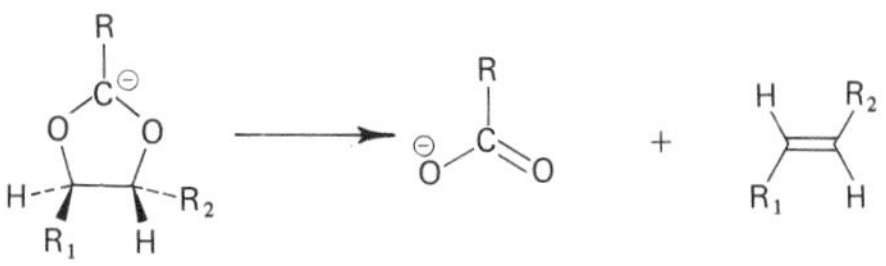

The cycloaddition reactions thus far discussed involve addition to pi-frameworks. It is therefore necessary to consider whether addition occurs to the same or to opposite faces of the pi-system (Fig. I.7). Woodward and Hoffmann have designated these addition modes *suprafacial* and *antarafacial* in accordance with

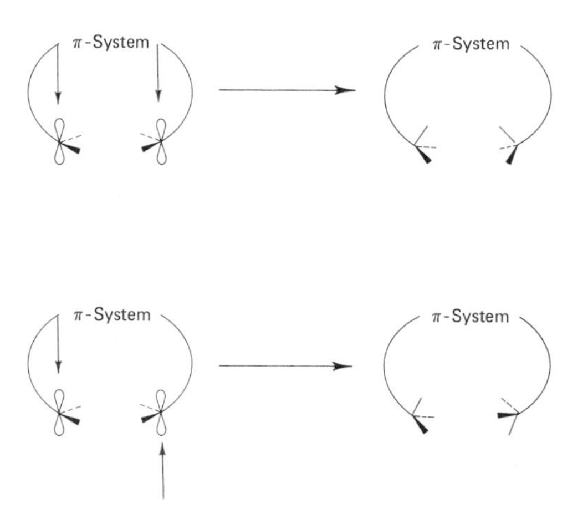

Fig. I.7. *Suprafacial and antarafacial addition to pi-systems.*

the terminology applied to sigmatropic reactions and to avoid the confusion that might result from the more conventional *cis* and *trans*. We shall adopt their convention. It should be noted that, in cycloaddition reactions, each component suffers addition, so the stereochemistry of ring formation at each pi-component must be specified. For example, extensive studies of the Diels-Alder reaction indicate that addition is suprafacial on each component [Eq. (I.10)].

(I.10)

The reaction may therefore be further described. This is accomplished by placing the subscript "a" (for antarafacial) or "s" (for suprafacial) after the number referring to the pi-component. That is, the Diels-Alder reaction is termed a $[4_s + 2_s]$ cycloaddition reaction.

Both "suprafacial" and "antarafacial" additions are well known, of course, for reactions involving noncyclic intermediates. There are distinct stereochemical consequences of each pathway, provided the reactant is appropriately labeled. Thus, for example, the (observed) antarafacial addition of bromine to *cis*-2-butene produces *dl*-2,3-dibromobutane, whereas the (unobserved) suprafacial addition of bromine to *cis*-2-butene would produce *meso*-2,3-dibromobutane (Fig. I.8).

There are also stereochemical consequences to suprafacial and antarafacial additions in cycloadditions reactions. For example, the cycloaddition $(\text{supra})_{\text{ethylene}}$–$(\text{supra})_{cis\text{-2-butene}}$ yields *cis*-1,2-dimethylcyclobutane, whereas the cycloaddition

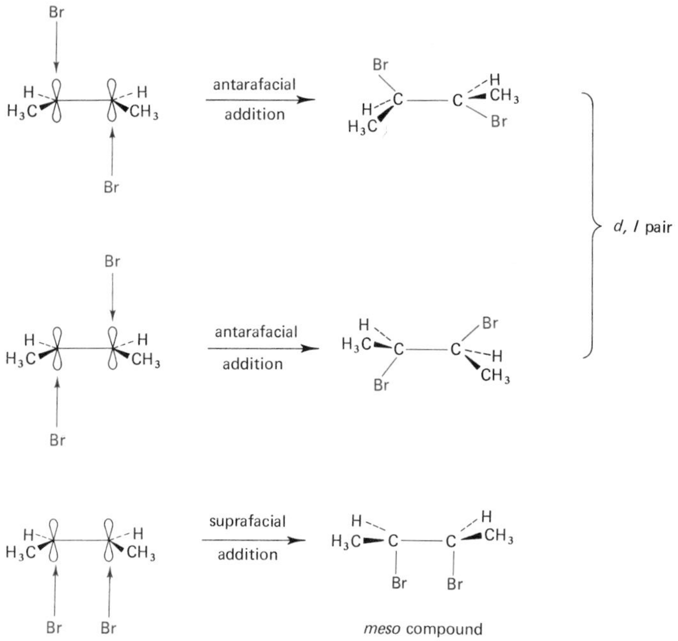

Br H H₃C H CH₃ Br antarafacial addition Br H H₃C C C H CH₃ Br

Fig. I.8. *"Suprafacial" and "antarafacial" additions of bromine to cis-2-butene. (The reader may wish to confirm that suprafacial addition to the other face of the system produces the same, meso, compound.)*

$(\text{supra})_{\text{ethylene}}$–$(\text{antara})_{cis\text{-2-butene}}$ yields *trans*-1,2-dimethylcyclobutane (Fig. I.9). Formally, it is possible to draw other $(\text{supra})_{\text{ethylene}}$–$(\text{antara})_{cis\text{-2-butene}}$ cycloaddition modes (Fig. I.10). However, the stereochemical result is in each instance identical: the originally *cis*-disposed methyl groups become *trans* in the product.

PROBLEM I.13

What are the products of

(a) $(\text{antarafacial})_{\text{ethylene}}$–$(\text{suprafacial})_{cis\text{-2-butene}}$ cycloaddition?

(b) $(\text{antarafacial})_{\text{ethylene}}$–$(\text{antarafacial})_{cis\text{-2-butene}}$ cycloaddition?

When two alkenes, each of which is appropriately labeled, undergo cycloaddition, the stereochemical possibilities expand.

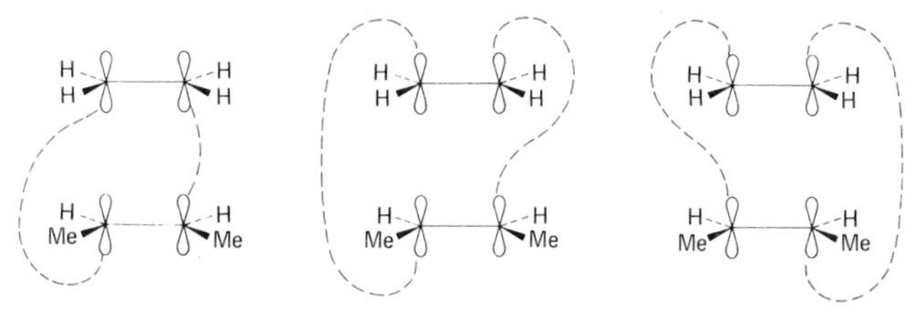

Fig. I.9. *Stereochemical consequences of* $(supra)_{ethylene}$*–*$(supra)_{cis\text{-}2\text{-}butene}$ *and* $(supra)_{ethylene}$*–*$(antara)_{cis\text{-}2\text{-}butene}$ *cycloaddition.*

Fig. I.10. *Other formal representations of the* $(supra)_{ethylene}$*–*$(antara)_{cis\text{-}2\text{-}butene}$ *cycloaddition.*

PROBLEM I.14

Consider the cycloaddition reaction of two molecules of *cis*-2-butene to produce 1,2,3,4-tetramethylcyclobutane. What are the products of

(a) (supra)–(supra) cycloaddition?

(b) (supra)–(antara) cycloaddition?

(c) (antara)–(antara) cycloaddition?

Even consideration of antarafacial modes of cycloaddition might be dismissed as a futile exercise, especially in $[2_s + 2_a]$ cycloadditions, since overlap considerations would appear to prohibit their operation. That would be a serious mistake. Electronic factors (controlled by molecular orbital symmetry) disfavor the $[2_s + 2_s]$ process and Woodward and Hoffmann have pointed out that an orthogonal, rather than "least motion," approach of reactants should be preferred (Fig. I.11). When the pi-frameworks are appreciably distorted, antarafacial addition becomes feasible. In rigid bi- and polycyclic molecules, furthermore,

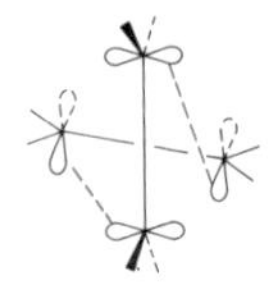

Fig. I.11. $[2_a + 2_s]$ *Cycloadditions of two ethylenes via orthogonal approach.*

the pi-components are sometimes held in geometries that uniquely favor antarafacial addition.

PROBLEM I.15

Classify the following as $[m + n + \cdots]$ cycloaddition reactions and indicate whether the addition is suprafacial or antarafacial on each component, if possible.

(a)

(b)

(c)

Cycloaddition reactions are also predicted to conform to a set of rules dependent upon the number of pi-electrons in each component and the mode of addition (Table I.5). We note that the Diels-Alder reaction (thermal, $[4_s + 2_s]$) conforms to the rules ($m + n = 6 = 4q + 2$) in Table I.5.

PROBLEM I.16

Review your classification of the cycloaddition reactions in Problems I.12 and I.15. Determine whether each should be thermally or photochemically allowed. *Assume suprafacial* addition in those cases where a stereochemical probe is lacking.

Table I.5 *Rules for [m + n] Cycloadditions*[a]

$m + n$	Allowed in ground state, forbidden in excited state	Allowed in excited state, forbidden in ground state
$4q$	$m_s + n_a$	$m_s + n_s$
	$m_a + n_s$	$m_a + n_a$
$4q + 2$	$m_s + n_s$	$m_s + n_a$
	$m_a + n_a$	$m_a + n_s$

[a] q = integer.

D. Pericyclic Reactions

Several individuals, notably Professors Dewar, Zimmerman, Fukui, and Woodward and Hoffmann, have observed that many concerted reactions involve transition states wherein all first-order bonding changes occur on a closed curve. Woodward and Hoffmann have termed such reactions *pericyclic* reactions and have shown that sigmatropic reactions, cycloaddition reactions, and electrocyclic reactions are all members of the pericyclic class of concerted reactions (Fig. I.12). Furthermore, they have demonstrated that it is possible to classify sigmatropic

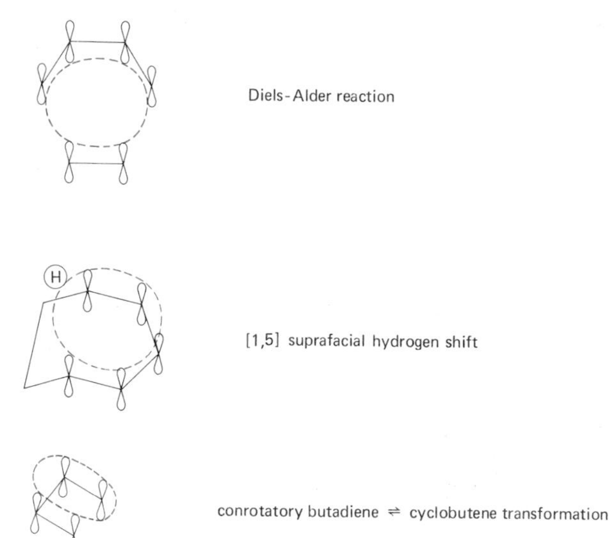

Fig. I.12. *Transition states for representative cycloaddition, sigmatropic, and electrocyclic reactions, showing closed-curve orbital interactions in the transition state.*

and electrocyclic reactions as cycloaddition reactions by examining the sigma-bond as a component in cycloaddition. The selection rules for cycloaddition reactions then apply to sigmatropic and electrocyclic reactions as well. Classification of the sense of involvement of the sigma-bond is determined as follows:

> Addition of the sigma-bond is *suprafacial* if either retention or inversion occurs at both termini of the bond.
> Addition of the sigma-bond is *antarafacial* if one terminus suffers inversion while the other suffers retention.

This classification of sigma-bonds as cycloaddition components arose from a consideration of suprafacial and antarafacial addition to double bonds, using the "ball and stick" model for the double bond (Fig. I.13). "Retention" involves

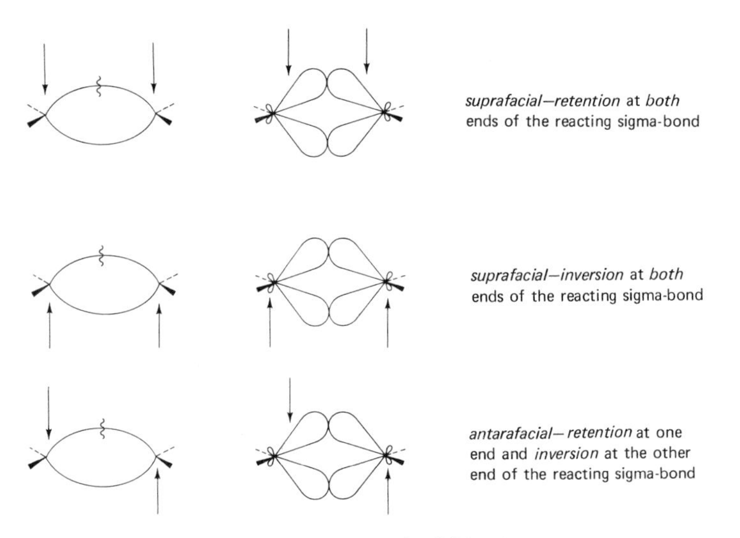

suprafacial—retention at *both* ends of the reacting sigma-bond
suprafacial—inversion at *both* ends of the reacting sigma-bond
antarafacial—retention at one end and *inversion* at the other end of the reacting sigma-bond

Fig. I.13. *Classification of sigma-bonds as cycloaddition components.*

interaction of the bonding lobe (large lobe) of the sigma-bond with the other cycloaddition component. "Inversion" involves interaction of the "tail" lobe with the other cycloaddition component. Thus, for the interactions shown for the conrotatory opening of cyclobutene, retention occurs at C_3 and inversion occurs at C_4 (Fig. I.14). The sigma-bond is thus involved in the antarafacial manner.

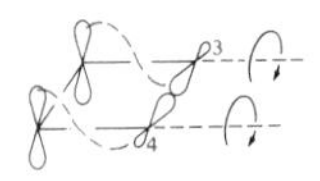

Fig. I.14. $[_{\sigma}2_a + _{\pi}2_s]$ *Cyclobutene* ⇌ *butadiene interconversion.*

The addition on the pi-bonds is then suprafacial, as shown, and the pathway may be classified as $[_{\sigma}2_a + _{\pi}2_s]$, where σ and π refer to the *types* of electrons involved.

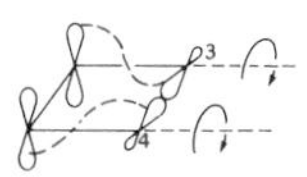

Fig. I.15. $[_{\sigma}2_s + _{\pi}2_a]$ *Cyclobutene* ⇌ *butadiene interconversion.*

It is important to note, however, that for precisely the same molecular motions described above, another classification is possible (Fig. I.15). In this instance, retention occurs at both ends of the sigma-bond, so it is involved in the suprafacial manner; however, addition on the pi-system is then antarafacial, as shown. This pathway may then be classified as $[_{\sigma}2_s + _{\pi}2_a]$. The reader will find that conrotatory opening will always lead to suprafacial involvement of one component and antarafacial involvement of the other. The selection rules for cycloaddition reactions (Table I.5) indicate that this mode of ring opening should be allowed (this is a $4q$ system, with one "a" and one "s" component). This prediction is thus precisely in accord with that following from the rules for electrocyclic reactions (Table I.1).

PROBLEM I.17

There is yet another $[_{\sigma}2_s + _{\pi}2_a]$ route for the above conrotatory opening. Show, formally, how this classification can be obtained.

PROBLEM I.18

Examine and classify the disrotatory opening of a cyclobutene according to the scheme just discussed. Consult Tables I.1 and I.5 to determine whether the predictions of the two approaches are in concordance.

PROBLEM I.19

Classify the conrotatory opening of a 1,3-cyclohexadiene to a 1,3,5-hexatriene according to this scheme. Consult Tables I.1 and I.5 to check your predictions.

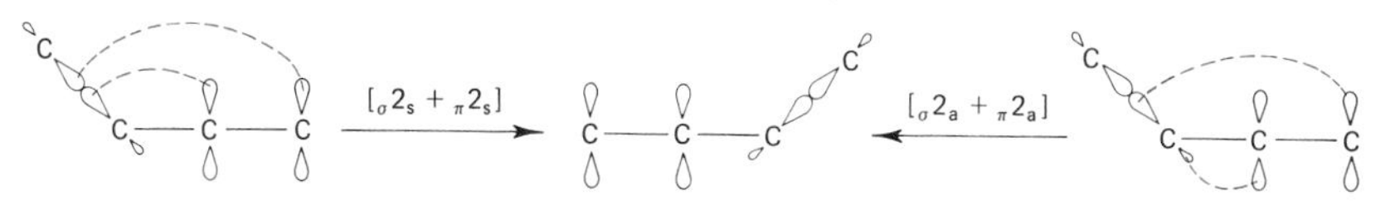

Fig. I.16. *Classification of a suprafacial [1,3] sigmatropic shift with retention at the migrating atom as* $[_{\sigma}2_s + _{\pi}2_s]$ *and* $[_{\sigma}2_a + _{\pi}2_a]$ *cycloaddition reactions.*

Sigmatropic reactions may be classified in similar fashion. For example, consider a [1,3] suprafacial shift with retention at the migrating atom (Fig. I.16). Two formal classifications for the reaction are possible. Most importantly, however, each of the two classifications leads to the same prediction—forbidden thermally, allowed photochemically (see Table I.5). Again, this is in accord with our previous conclusions.

PROBLEM I.20

Analyze a suprafacial [1,3] sigmatropic shift with *inversion* at the migrating atom by the above scheme. Is it allowed or forbidden thermally according to the rules presented in Table I.5?

PROBLEM I.21

Classify a suprafacial [1,5] shift with *retention* at the migrating atom by the same method. Is it allowed or forbidden thermally according to the rules?

Woodward and Hoffmann have generalized the rules for two-component cycloaddition reactions (Table I.5) to include multicomponent reactions. In that case, the selection rule for all pericyclic reactions is as follows:

> A ground-state pericyclic change is symmetry-allowed when the total number of $(4q + 2)_s$ and $(4r)_a$ components is odd.

If the total is even, the reaction is forbidden thermally, but allowed photochemically. For example, suppose that a reaction were classified as $[{}_{\pi}6_a + {}_{\sigma}2_a + {}_{\sigma}2_s]$. Neither ${}_{\pi}6_a$ nor ${}_{\sigma}2_a$ can be categorized as a $(4q + 2)_s$ or a $(4r)_a$ component [they are both $(4q + 2)_a$ components], and they are not counted. However, ${}_{\sigma}2_s$ is a $(4q + 2)_s$ component and is counted. Thus, in this instance, the total number of $(4q + 2)_s$ and $(4r)_a$ components is 1, an odd number. The reaction is therefore thermally allowed.

PROBLEM I.22

Reexamine the reactions in Problems I.8 and I.1 and classify them according to the above method. Determine which are thermally and which are photochemically allowed and compare your predictions with those previously obtained.

The reader will find (Problem I.22) that the predictions made using the new rules are precisely those that derived from the previous analyses. It is significant, however, that the above approach to describing concerted reactions is the most general and can be applied to all types of pericyclic concerted reactions. Thus, consider the following reaction [Eq. (I.11)]:

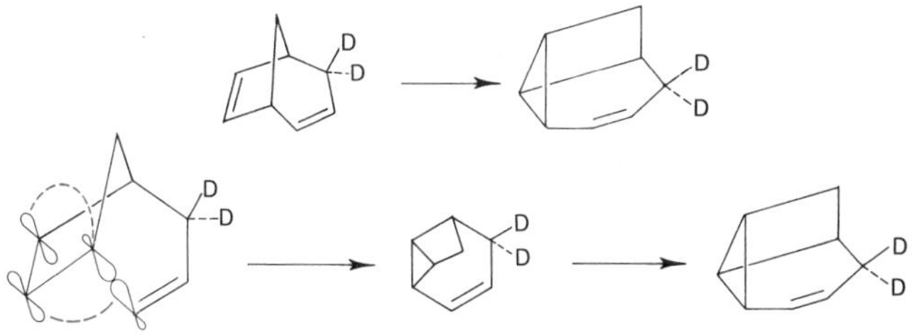

(I.11)

The reaction may be classified as a $[_{\sigma}2_a + {_\pi}2_a]$ cycloaddition (the reader may wish to consult models). Since the total $(4q + 2)_s + (4r)_a = 0$, the reaction is forbidden thermally and allowed photochemically. Note that, in this instance, the rigid geometry of the bicyclic system forces the reaction (if concerted) to be antarafacial on both components.

PROBLEM I.23

Classify the following reaction. The implicated bonds are shown (the use of models is advised). If concerted, would the reaction be allowed thermally or photochemically?

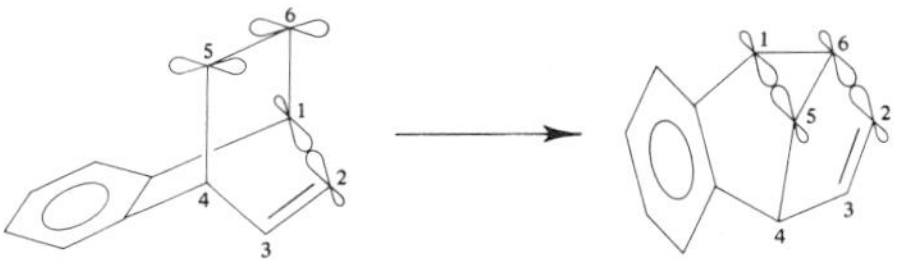

PROBLEM I.24

Classify the reaction shown below. Should it be allowed thermally or photochemically?

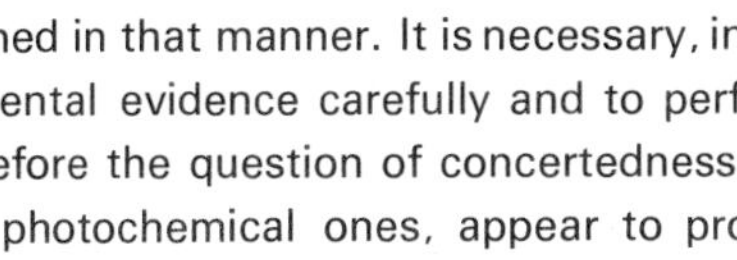

We wish, at this time, to offer some words of caution. The formation of a product that can result formally from an "allowed" pathway does not demand that it actually be formed in that manner. It is necessary, in each case, to examine the available experimental evidence carefully and to perform additional experiments, if required, before the question of concertedness can be settled. Many reactions, especially photochemical ones, appear to proceed nonconcertedly, although the products are those predicted from the rules. Also, the generalized rules for pericyclic reactions should be applied with care. In some instances, it is found that special geometrical factors lead to the imposition of energy barriers not foreseen by rigid application of the rules (when multicomponent reactions are involved). It is then advisable to examine the reaction in the context of the theories to be presented in the next chapter.

Chapter II

THEORY

The number of papers describing theoretical approaches to concerted reactions are now legion. They are almost all based ultimately upon molecular orbital theory, but each emphasizes particular approaches within the area: the construction of correlation diagrams, frontier orbital theory, perturbational molecular orbital theory, etc. In this chapter, we shall discuss aspects of these methods and provide examples of the application of each. Each of the methods has been discussed in detail by its proponent, and references to the original work are included at the end of this chapter.

Essentially, it is necessary to devise a way of determining the relative transition-state energies for the particular stereochemical modes of transformation we wish to compare. For example, we shall need to determine whether the orbital interactions in the transition state for the conrotatory closure of 1,3,5-hexatriene to 1,3-cyclohexadiene are more, or less, favorable than the orbital interactions for disrotatory closure (Fig. II.1). Since both reactions proceed via a common reactant, an assessment of the relative transition-state energies will enable us to determine which reaction should proceed more rapidly.

Two approaches have been applied toward determining the relative transition-state energies. One of them examines the energy changes which result from

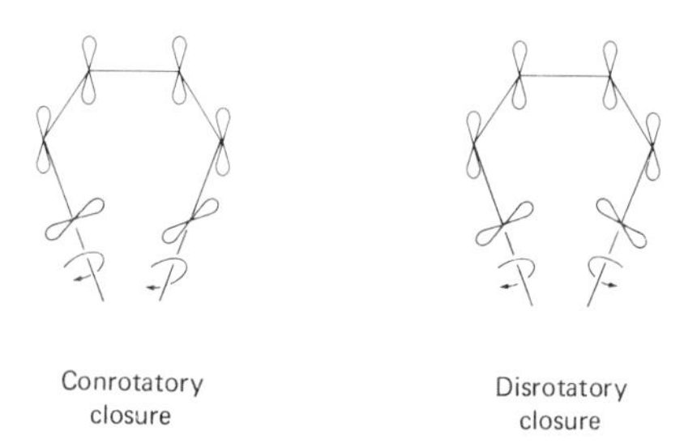

Fig. II.1. *Orbital interactions in the transition state for conrotatory and disrotatory closure of hexatriene.*

interactions of molecular orbitals as the reactant is converted to the product. A knowledge of those energy changes indirectly affords information about the energy of the transition state since it must lie at a position between reactant and product on the reaction coordinate. The other approach examines the transition state of a given concerted reaction *directly* and determines whether it is stabilized or destabilized by electronic factors. We shall begin our discussion first by considering those approaches which indirectly examine the transition-state energy.

A. Correlation Diagrams

As an example we choose the cyclobutene ⇌ butadiene interconversion. The first step of our analysis entails recognition of the bonds intimately involved in the process [Eq. (II.1)].

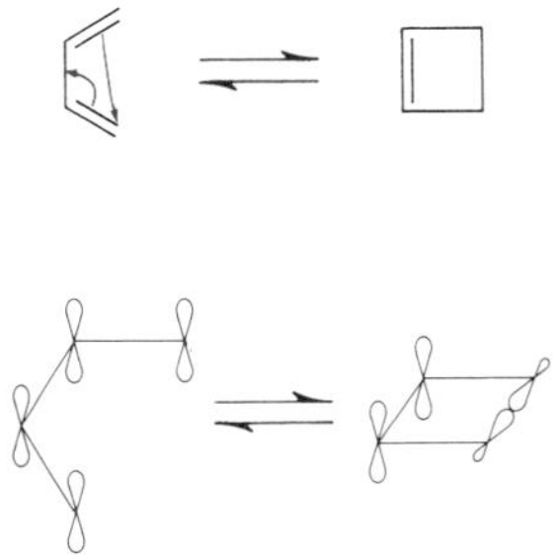

(II.1)

Clearly, this conversion involves transformation of the four-pi-electron butadiene system to a two-pi-electron, two-sigma-electron cyclobutene system. We next establish molecular orbitals (MO's) for the reactant and product systems (Fig. II.2). At this stage, the crucial function of orbital symmetry emerges. It will be noted that each lobe of an atomic orbital in Fig. II.2 has an algebraic sign. The *relative phases* of two interacting lobes (i.e., their relative algebraic signs) will determine whether that interaction is stabilizing or destabilizing. Indeed, the importance of relative phases is apparent in the energy-ordering of the molecular orbitals of reactant and product. Thus, the most stable butadiene molecular orbital is that in which all lobes on the same face have the same phase (ψ_1), a situation which occurs because overlap of orbitals with like phase (sign of coefficient) is a favorable interaction. Electron density is concentrated between the nuclei of the atoms involved, affording a *bonding* interaction. On the other hand, overlap of orbitals of *unlike* phase is an unfavorable, *antibonding* interaction. It tends to keep electron density away from the region between the nuclei. We note that the butadiene orbitals increase in energy in the order ψ_1, ψ_2, ψ_3, ψ_4 as the number of

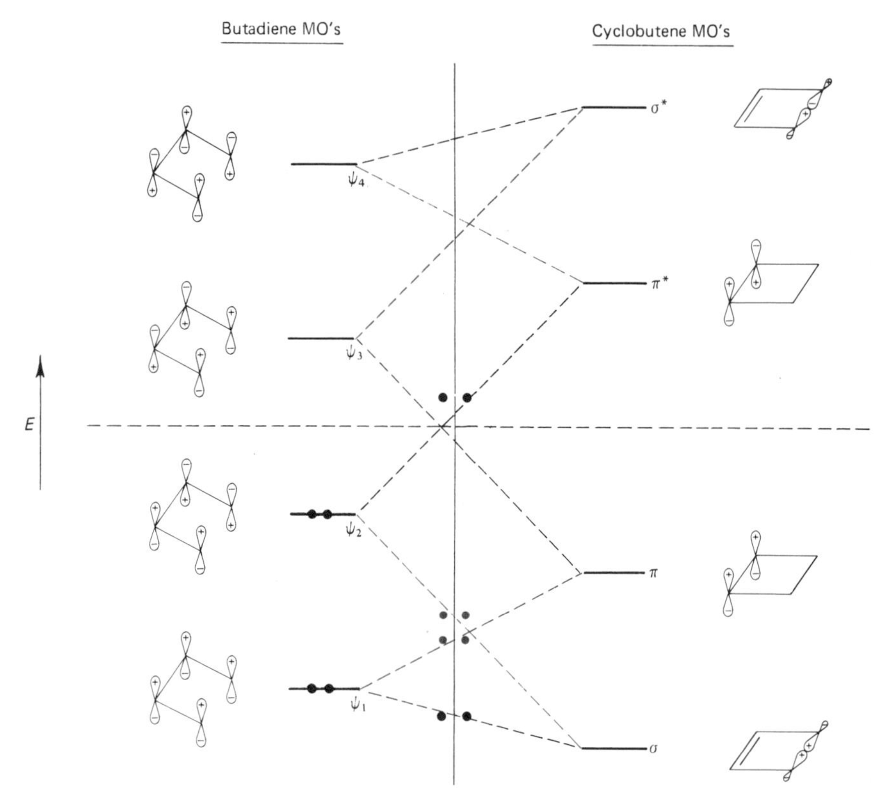

Fig. II.2. *Correlation diagrams for butadiene ⇌ cyclobutene interconversions.*

antibonding interactions increases. The orbitals we cite here can be derived by using simple Hückel theory.*

The transition state will occur somewhere along the reaction coordinate between reactant and product. In Fig. II.2, a vertical line has been drawn (at an arbitrary position) to locate the transition state. The precise location of that line is not crucial to the arguments that follow. The energy of the transition state will depend critically upon the energy changes experienced by individual orbitals as they transform from reactant to product MO's. In Fig. II.2 are shown two possible orbital correlations, one in black and one in red. We shall see, later, that one correlation corresponds to the conrotatory and the other to the disrotatory motion of the terminal orbitals. At this time, we shall refer to them simply as the "red pathway" and the "black pathway." The orbital correlations implicit in Fig. II.2 are listed in Table II.1.

*Those unfamiliar with simple Hückel theory and the construction of correlation diagrams are referred to Appendixes A and B. Appendix B would be most profitably consulted after reading this chapter.

Table II.1 *Orbital Correlations Obtained from Fig. II.1*

Red pathway[a]	Black pathway[a]
$\psi_1 \longleftrightarrow \pi$	$\psi_1 \longleftrightarrow \sigma$
$\psi_2 \longleftrightarrow \sigma$	$\psi_2 \longleftrightarrow \pi^*$
$\psi_3 \longleftrightarrow \sigma^*$	$\psi_3 \longleftrightarrow \pi$
$\psi_4 \longleftrightarrow \pi^*$	$\psi_4 \longleftrightarrow \sigma^*$

[a]We adopt the convention that a double-headed arrow ($\longleftrightarrow$) signifies "correlates with."

Let us examine the effect upon transition-state energy of the correlations shown for the *thermal* interconversion of butadiene to cyclobutene. There are four electrons involved in the process and they will occupy, in pairs, the two lowest MO's of butadiene, ψ_1 and ψ_2. We can describe the electronic ground state of butadiene as $\psi_1^2\psi_2^2$, which indicates that two electrons occupy ψ_1 and two occupy ψ_2. If the correlation observed is that indicated by the red lines, the electronic energy at the transition state will be determined by the intersection of the vertical line with the dashed lines emanating from ψ_1 and ψ_2. Similarly, if the black pathway is followed, the electronic energy of the transition state will be determined by intersection of the vertical line with the black dashed lines emanating from ψ_1 and ψ_2. These points are indicated in Fig. II.1 by pairs of red and black dots. It is important to note that these transition-state energies are quite different and that, in this case, the energy at the transition state will surely be lower for the process leading to the red correlation. That process is then said to be *allowed*, while the higher-energy black pathway is *forbidden*.

Another way of reaching the same conclusion derives from a consideration of the *electronic states* of reactant and product that correlate in the two pathways. To do this, refer to Table II.1, which indicates how each orbital correlates. In the black pathway, $\psi_1 \longleftrightarrow \sigma$ and $\psi_2 \longleftrightarrow \pi^*$, so that the ground electronic state of butadiene $\psi_1^2\psi_1^2$ correlates with a much higher energy doubly excited state of cyclobutene ($\sigma^2\pi^{*2}$). This attempt to correlate a ground-state molecule with a much higher energy molecule will necessarily be thwarted by imposition of a large energy barrier in the transition state. The reaction is *forbidden*.

In the red pathway, $\psi_1 \longleftrightarrow \pi$ and $\psi_2 \longleftrightarrow \sigma$, so that the ground state of butadiene, $\psi_1^2\psi_2^2$, correlates with the ground state of cyclobutene ($\sigma^2\pi^2$). There is no special electronic energy barrier imposed upon the reaction, and it is *allowed*. In general, when such a correlation diagram can be drawn, pathways that lead to interconversion of ground electronic states are *symmetry allowed* (thermally!) and those that correlate a ground state with an electronically excited state are *symmetry forbidden*. In similar fashion, for a photochemical reaction proceeding via the first excited state to be allowed, first excited states of reactant and product must smoothly correlate.

PROBLEM II.1

In the black pathway described above, with what electronic state of butadiene does the ground electronic state of cyclobutene correlate? How would you describe that state?

PROBLEM II.2

What are the first excited photochemical states of butadiene and cyclobutene? (They are formed by promoting one electron from the highest occupied molecular orbital of the ground state to the lowest vacant molecular orbital of the ground state in each case.) In which case, the red or the black pathway, do they correlate? What are the intended correlations in the other pathway?

The reader who worked Problem II.2 will have discovered that, although only the red pathway led to a correlation of ground states, only the black pathway led to a correlation of first excited states. That is, one process is allowed thermally but forbidden photochemically (red), whereas the other is forbidden thermally but allowed photochemically (black). That is precisely the type of behavior we hoped to explain and we see that it can be said to result from different correlations of molecular orbitals that result from different pathways.

But, why are these particular correlations observed, why are they different, and which of the pathways, red or black, corresponds to disrotatory and which corresponds to conrotatory motion of the terminal orbitals? It is essential to realize that the conrotatory and disrotatory processes differ fundamentally. In Fig. II.3 are shown end views of the terminal orbitals during conrotatory and disrotatory closure. The conrotatory mode is characterized by a C_2 axis of

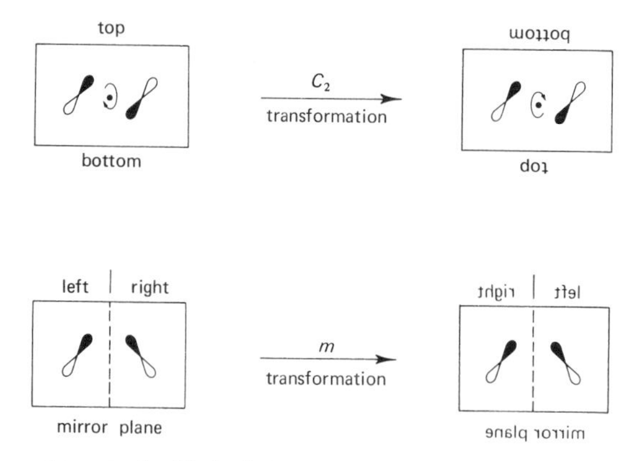

Fig. II.3. *View of terminal orbitals during conrotatory and disrotatory closure.*

symmetry (rotation through 180°) whereas the disrotatory process is characterized by a plane of symmetry. Those symmetry operations will transform orbital lobes as shown in Fig. II.3 (black to black, white to white). Now, the molecular orbitals of butadiene and cyclobutene themselves have a C_2 axis and a plane as symmetry elements. Each molecular orbital can be classified as symmetric (S) or antisymmetric (A) with respect to the symmetry element if operation of the symmetry element leads to an unchanged molecular orbital or a new molecular orbital in which all signs have been reversed, respectively.

For example, let us examine ψ_1 of butadiene (Fig. II.4). We examine the *s-cis* conformation, which is required for ring closure to a cyclobutene. The molecule

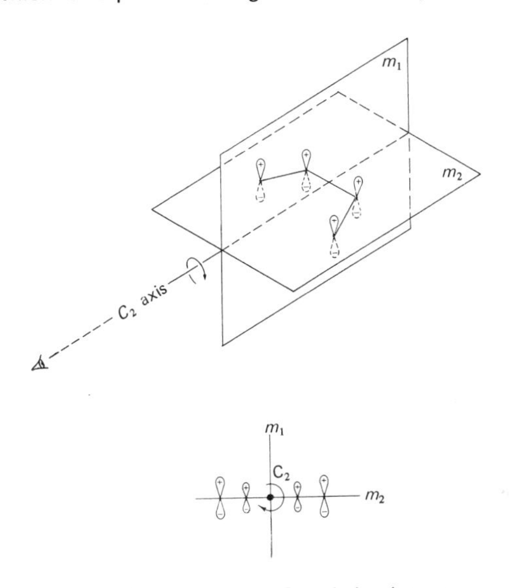

Fig. II.4. *View of butadiene ψ_1 orbital along C_2 axis in plane m_2.*

lies in plane m_2 and is bisected by the reflection plane m_1. Also, the C_2 axis is indicated in the diagram. It should be clear that reflection through plane m_1 merely converts the MO into itself (no signs are changed). Thus, ψ_1 is symmetric (S) with respect to m_1. Also, it is evident that rotation through 180° (C_2) will reverse the sign of each orbital lobe. Thus, ψ_1 is antisymmetric (A) with respect to the C_2 operation. Similarly, each molecular orbital of butadiene and cyclobutene can be classified as symmetric or antisymmetric with respect to symmetry operations C_2 and m_1. This is illustrated for the disrotatory conversion of butadiene to cyclobutene in Fig. II.5. Since a symmetry plane characterizes the disrotatory process, each orbital is classified as symmetric or antisymmetric with respect to that symmetry element, as shown. The reader should verify the assignments in Fig. II.5. Next, since the disrotatory process itself is characterized by the same plane of symmetry, an orbital must maintain its symmetry (S or A) as it is transformed into a product

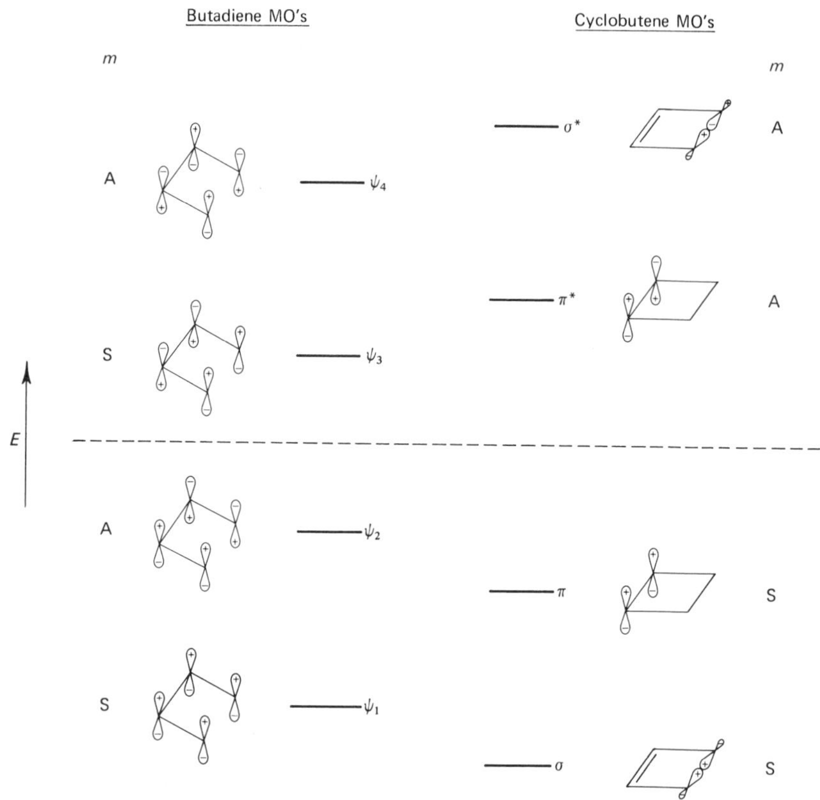

Fig. II.5. *Classification of butadiene and cyclobutene MO's for disrotatory closure.*

orbital. That is, an S orbital of reactant will correlate with an S orbital of product, or an A orbital of reactant will correlate with an A orbital of product, but an A reactant orbital cannot correlate with an S product orbital, or *vice versa*. As the final step in construction of the correlation diagram, orbitals of like symmetry are joined, with observance of the noncrossing rule, which states that orbitals of like symmetry will not cross due to electron repulsion. The completed correlation diagram is shown in Fig. II.6. The reader will note that this is precisely the "black correlation" indicated in Fig. II.2, so we may identify the "black correlation" with disrotatory motion. The conclusions reached about this process are unchanged: The reaction should be forbidden thermally and allowed photochemically if it proceeds in the disrotatory fashion.

Similarly, we may construct a correlation diagram for the conrotatory process, except this time the orbitals are classified with respect to C_2 (Fig. II.7). The reader will note that this correlation is precisely that observed in the red pathway of Fig. II.2. The conrotatory process is allowed thermally and forbidden photochemically. The reader may still wonder *why* the energies of some orbitals

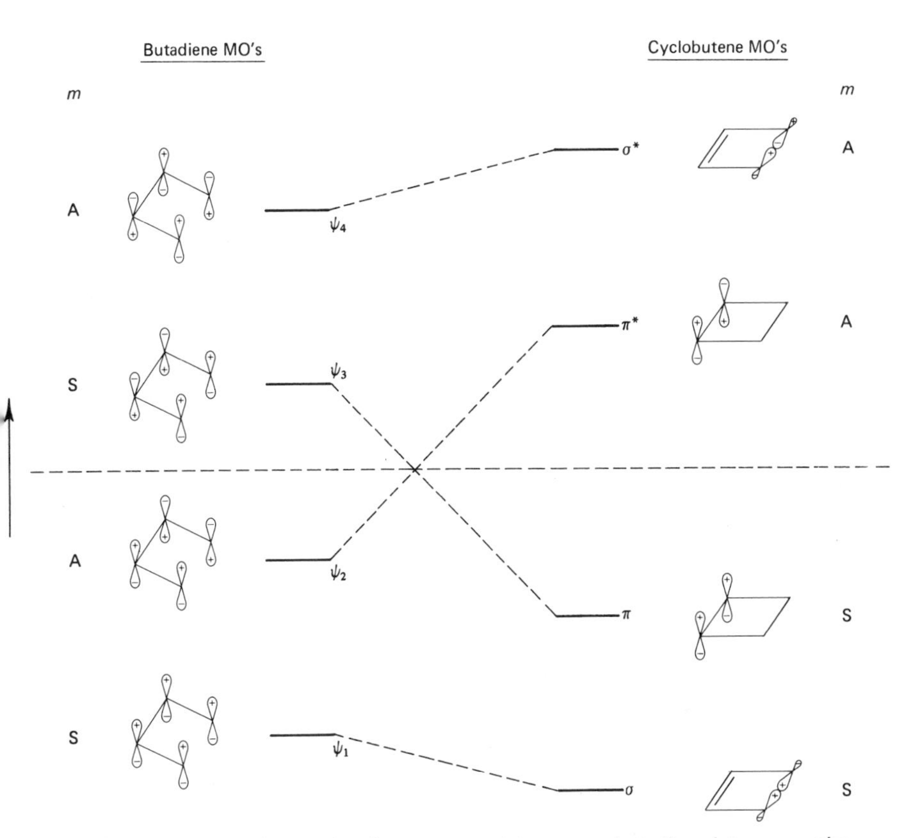

Fig. II.6. *Correlation diagram for disrotatory cyclobutene ⇌ butadiene interconversion.*

increase (upward slope) while others decrease (downward slope) as the reaction proceeds along the reaction coordinate. That fact may be ascertained by an examination of the terminal orbitals during cyclization. Thus, we examine the terminal lobes during conrotatory closure of ψ_1 and ψ_2 (Fig. II.8). For ψ_1, as cyclization proceeds an *antibonding* interaction develops at the terminal lobes, with consequent energy increase (upward slope) with progress along the reaction coordinate. For ψ_2 the interaction at the terminal lobes is bonding, with consequent energy decrease as the reaction proceeds. Similarly, ψ_3 will increase in energy and ψ_4 will decrease in energy as cyclization occurs. It is gratifying to find that the results of the mechanical classification of orbitals with respect to symmetry elements, and their correlation thereby, is in accord with expectations based upon an examination of bonding interactions during the reaction.

PROBLEM II.3

Draw the terminal orbital interactions for $\psi_1 - \psi_4$ during *disrotatory* closure, and indicate whether they are bonding or antibonding.

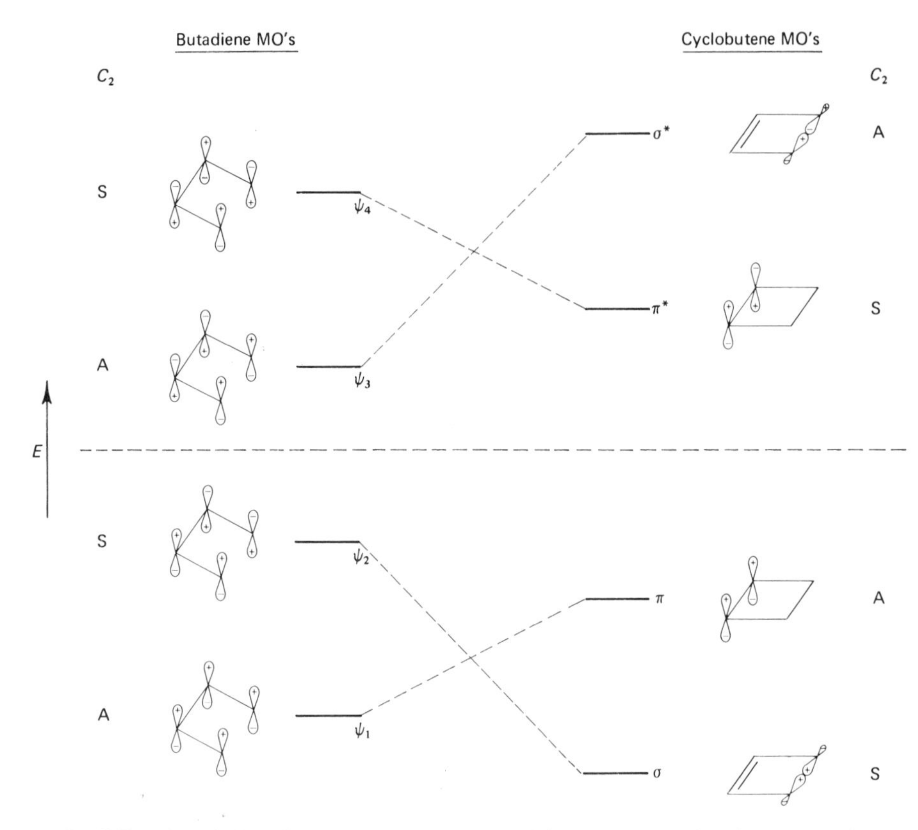

Fig. II.7. *Correlation diagram for conrotatory cyclobutene ⇌ butadiene interconversion.*

The arguments in the previous paragraph allow one to understand why unsymmetrical 1,3-dienes such as 2-methyl-1,3-butadiene (isoprene) should follow the same course of cyclization as the corresponding symmetrical diene (butadiene). Strictly speaking, the added methyl group renders classification of MO's as S or A untenable, since the coefficients of the orbitals at the formerly identical carbon atoms (C_1, C_4 and C_2, C_3) are no longer of equal magnitude. However, so long as the relative *signs* of the coefficients of the MO's are not affected by the perturbing group, as is almost always the case, the initial upward and downward slopes will be almost identical for the symmetrical and unsymmetrical systems.

Fig. II.8. *Bonding and antibonding interactions at the terminal lobes of butadiene orbitals during conrotatory transformation.*

This situation arises since the fundamental bonding and antibonding interactions will be virtually the same. As a result, the transition-state energies are likely to be very similar for the symmetrical and unsymmetrical systems.

PROBLEM II.4

Given the molecular orbitals below, construct correlation diagrams for the conrotatory and disrotatory conversion of allyl to cyclopropyl (or the reverse process).

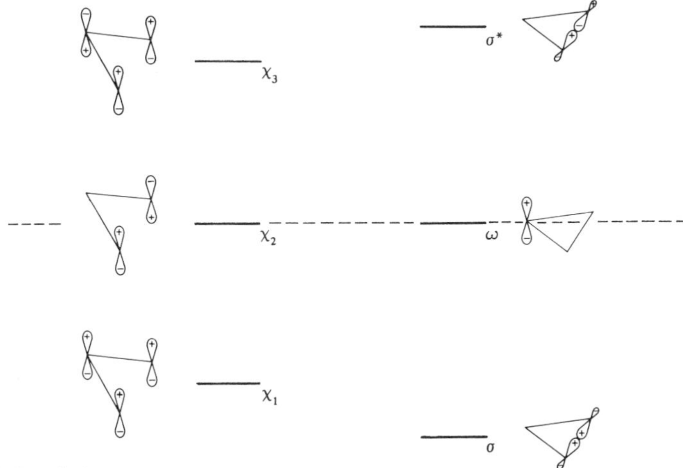

PROBLEM II.5

For the thermal allyl ⇌ cyclopropyl *carbonium ion* interconversion, determine which pathway is symmetry allowed and which is symmetry forbidden. Also, how should the first excited states of the carbonium ion interconvert?

PROBLEM II.6

For the thermal allyl ⇌ cyclopropyl anion interconversion, determine which pathway is symmetry allowed and which is symmetry forbidden.

PROBLEM II.7

What is the electronic ground state of allyl radical? Of cyclopropyl radical? Which process (conrotatory or disrotatory) will interconvert them?

The device of correlation diagrams can also be easily applied to cycloaddition reactions in which a suitable amount of symmetry is present. For example, let us consider the classic $[_{\pi}4_s + _{\pi}2_s]$ reaction. The reaction is imagined to proceed via approach, in two separate planes, of the diene and dienophile, as shown in Fig. II.9. An end view of the interaction of butadiene and ethylene at fairly large

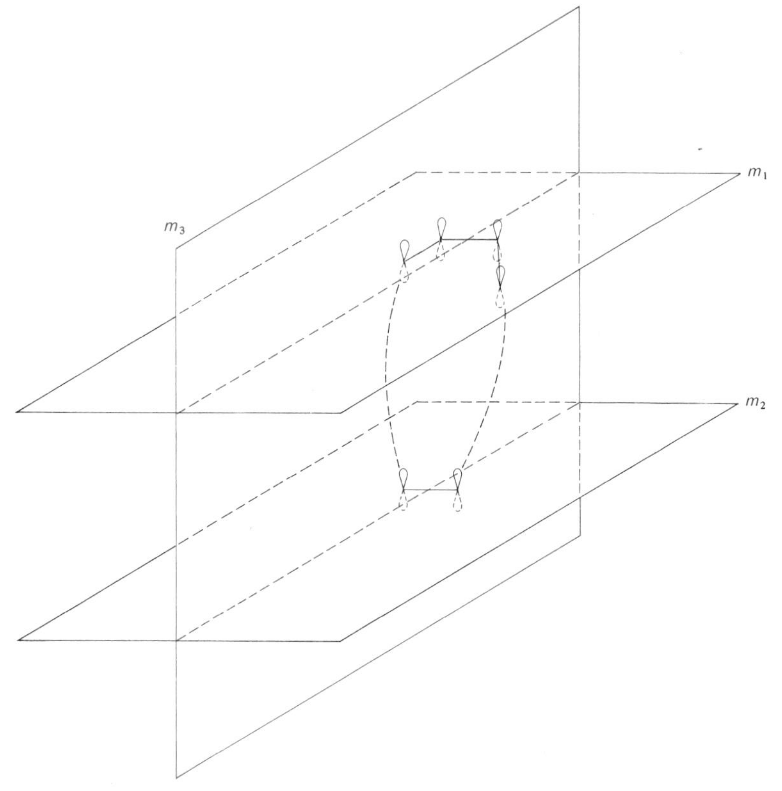

Fig. II.9. *Geometry of approach of butadiene and ethylene in the Diels-Alder* $[_{\pi}4_s + _{\pi}2_s]$ *cycloaddition.*

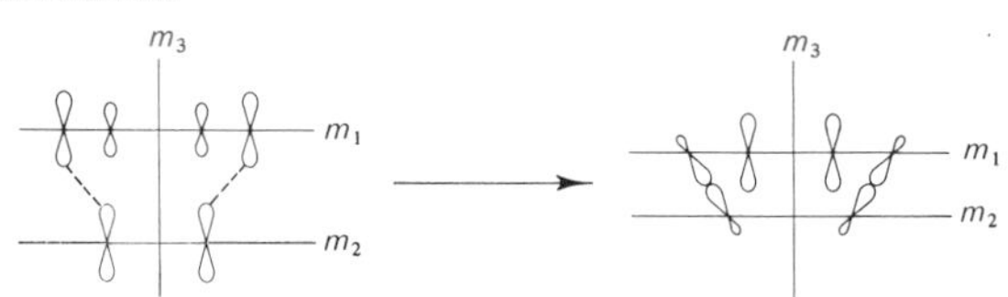

Fig. II.10. *View along planes m_1 and m_2 of orbitals at beginning and termination of interaction.*

distances and at the termination of that interaction by formation of the product, cyclohexene, is shown in Fig. II.10. One seeks to find the maximum number of symmetry elements that characterize the reaction as the molecules approach, bonding occurs, and the product is formed. The reader will note that the only symmetry element present throughout the reaction is the symmetry plane, m_3. It is then necessary to generate MO's which will be either symmetric (S) or antisymmetric (A) with respect to operation of that symmetry element. This is accomplished in Fig. II.11. Since each shares m_3 as a symmetry element, it is sufficient to use the ethylene and butadiene MO's at large distances of approach,

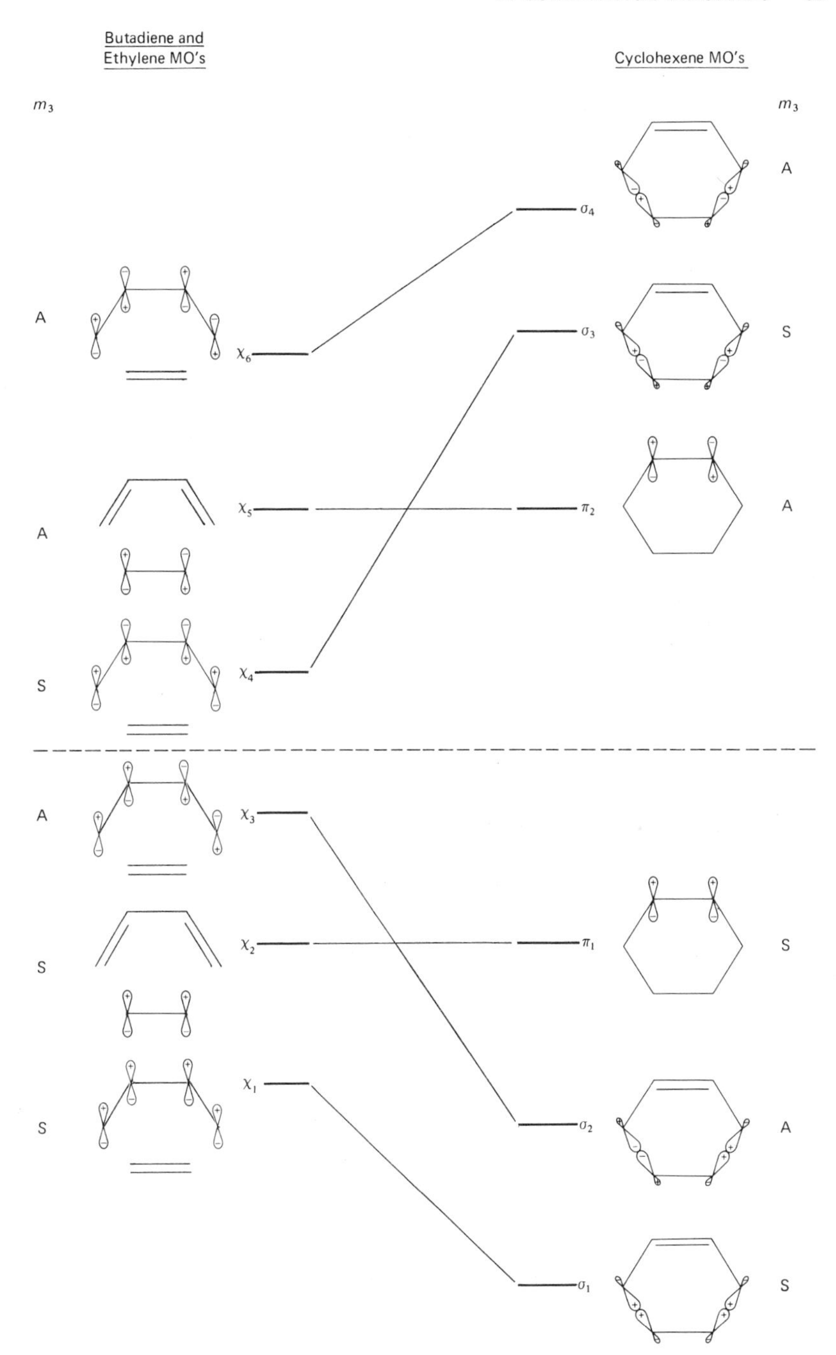

Fig. II.11. *Correlation diagram for the Diels-Alder* $[_{\pi}4_s + _{\pi}2_s]$ *reaction.*

as is shown in Fig. II.11. On the other hand, the MO's of the newly formed sigma-bonds in cyclohexene must be generated as combinations of the individual bonding and antibonding sigma-bonds, as shown on the right-hand side of Fig. II.11. It is then possible to classify each MO as symmetric or antisymmetric with respect to m_3, as is shown in Fig. II.11. The reaction will be thermally allowed if ground electronic states correlate and forbidden otherwise. The reaction will be photochemically allowed if first excited states correlate, and forbidden otherwise. The orbital correlations obtained from Fig. II.11 are shown in Table II.2. Six electrons are involved in the transformation, so there is a smooth correlation of ground states ($\chi_1^2\chi_2^2\chi_3^2 \longleftrightarrow \sigma_1^2\sigma_2^2\pi_1^2$). The reaction is *thermally allowed*. On the other hand, the photochemical reaction is *forbidden* to proceed concertedly, since the first excited states, $\chi_1^2\chi_2^2\chi_3^1\chi_4^1$ and $\sigma_1^2\sigma_2^2\pi_1^1\pi_2^1$, do not correlate. Rather, each correlates with a much higher energy state ($\chi_1^2\chi_2^2\chi_3^1\chi_4^1 \longleftrightarrow \sigma_1^2\sigma_2^1\pi_1^2\sigma_3^1$ and $\sigma_1^2\sigma_2^2\pi_1^1\pi_2^1 \longleftrightarrow \chi_1^2\chi_2^1\chi_3^2\chi_5^1$) with a consequent symmetry-imposed barrier to reaction.

Table II.2 *Orbital Correlations for the Diels-Alder Reaction Obtained from Fig. II.11*

$\chi_1 \longleftrightarrow$	σ_1
$\chi_2 \longleftrightarrow$	π_1
$\chi_3 \longleftrightarrow$	σ_2
$\chi_4 \longleftrightarrow$	σ_3
$\chi_5 \longleftrightarrow$	π_2
$\chi_6 \longleftrightarrow$	σ_4

PROBLEM II.8

Given the orbitals below (next page), draw a correlation diagram for the suprafacial–suprafacial addition of pentadienyl to ethylene.

PROBLEM II.9

Using the MO's of Problem II.8, determine whether each of the following reactions is allowed thermally or photochemically:

(a) Pentadienyl anion + ethylene $\longrightarrow$ cycloheptenyl anion

(b) Pentadienyl cation + ethylene $\longrightarrow$ cycloheptenyl cation

B. Frontier Orbital and Related Methods

In our discussion of correlation diagrams, we were careful to consider the energy change of each orbital as it transformed from reactant to product. Nevertheless, it has been found that, in many instances, it is possible to reach the same

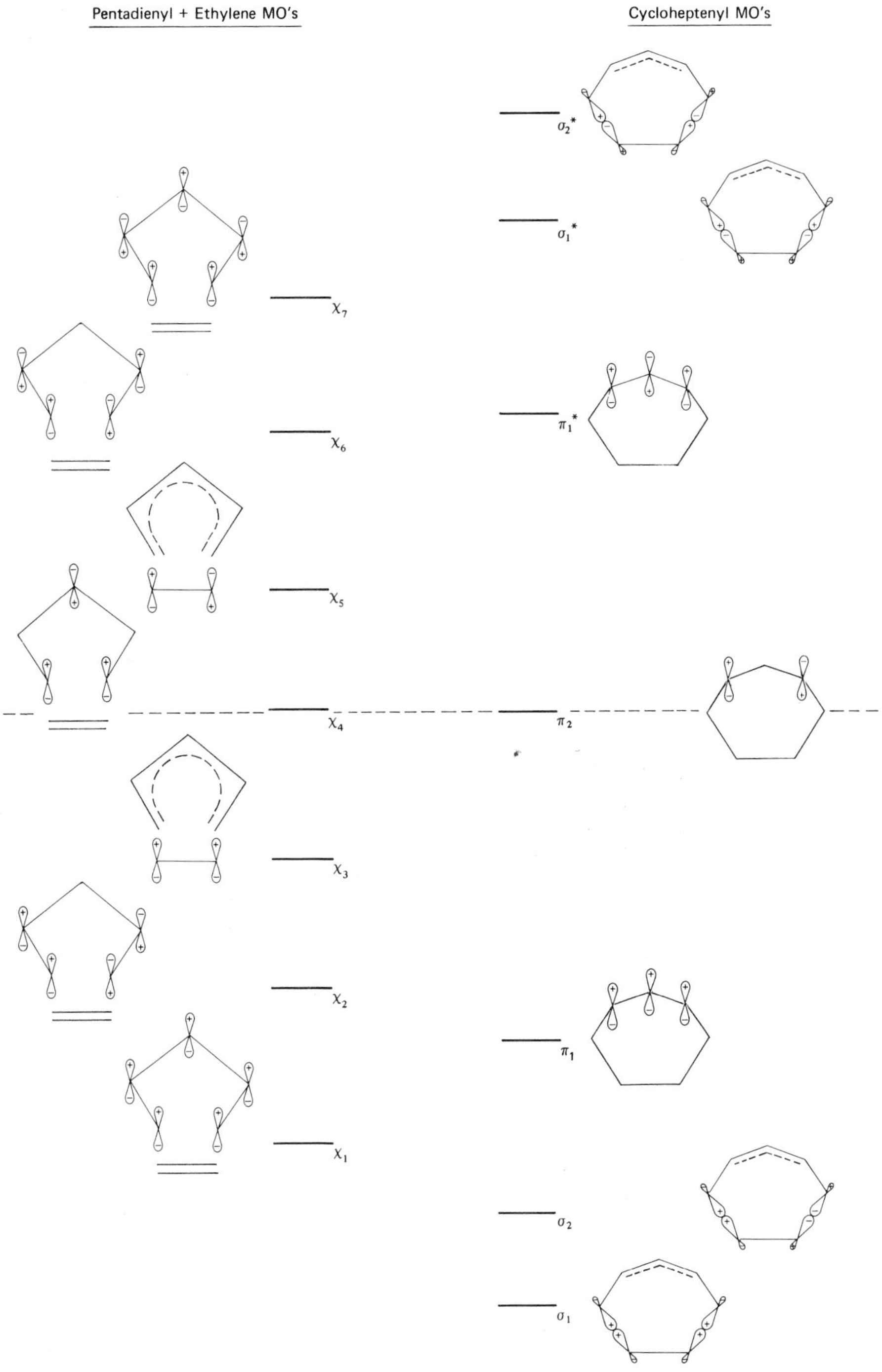

Pentadienyl + Ethylene MO's

Cycloheptenyl MO's

σ_2^*

σ_1^*

χ_7

π_1^*

χ_6

χ_5

χ_4

π_2

χ_3

χ_2

π_1

χ_1

σ_2

σ_1

conclusions by an examination of select, *frontier* orbitals. These orbitals, the highest occupied molecular orbital (HOMO) and the lowest vacant molecular orbital (LVMO), often contribute most to the overall energy change as a transformation occurs. This, and related methods have been applied to cycloaddition and electrocyclic and sigmatropic reactions as well as *exo-endo* relationships by Fukui and by Woodward and Hoffmann.

First, let us consider the Diels-Alder reaction, the $[_{\pi}4_s + _{\pi}2_s]$ cycloaddition of ethylene to butadiene. The signs of the coefficients of the molecular orbitals of each and the occupancy of the orbitals in the thermal reaction are shown in Fig. II.12. The HOMO's and LVMO's are labeled in Fig. II.12. In the frontier orbital approach, the most significant interactions are considered to be between the HOMO of one component and the LVMO of the other. In the Diels-Alder reaction, the interactions are favorable (Fig. II.13). It should be emphasized that consideration of these orbitals alone certainly is an approximation. The approximation will be best when the relevant HOMO's and LVMO's are close in energy.

Electrocyclic reactions may also be treated from the frontier orbital approach. Fukui, using a perturbation method, has shown that transformation of the cyclic

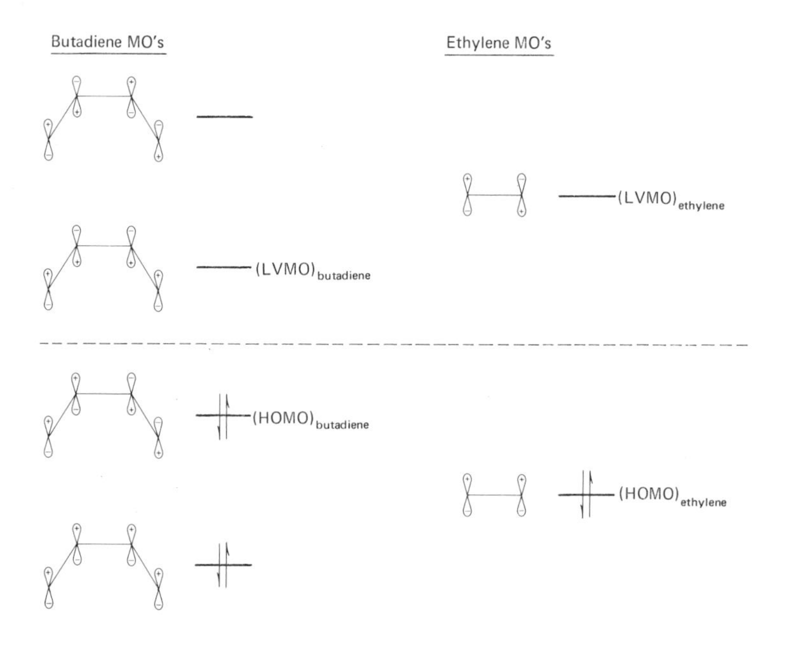

Fig. II.12. *Butadiene and ethylene MO's.*

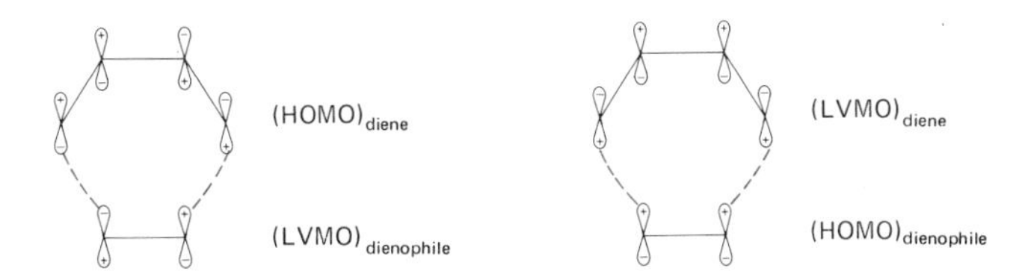

Fig. II.13. *HOMO–LVMO interactions in the Diels-Alder reaction.*

structure to the acyclic polyene may be considered as a cycloaddition of the sigma-bond to the pi-system of the cyclic molecules [Eq. (II.2)]:

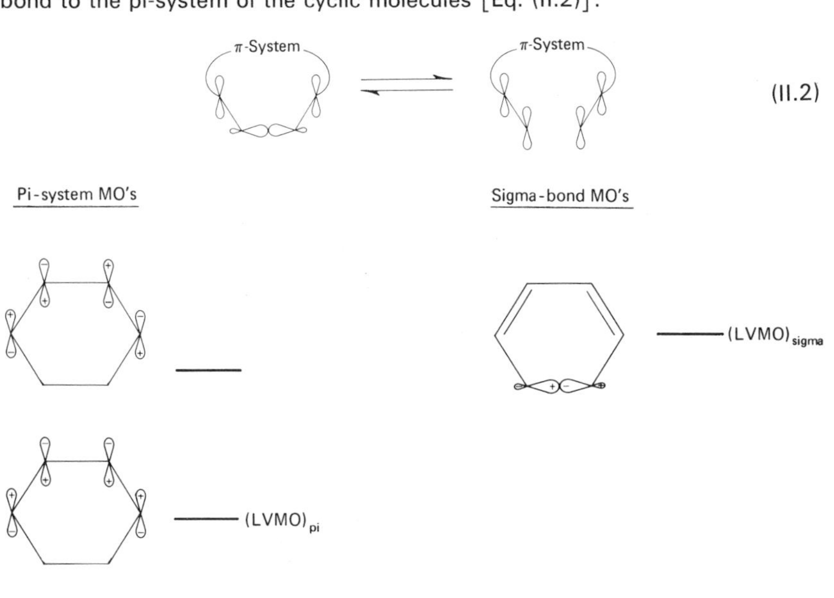

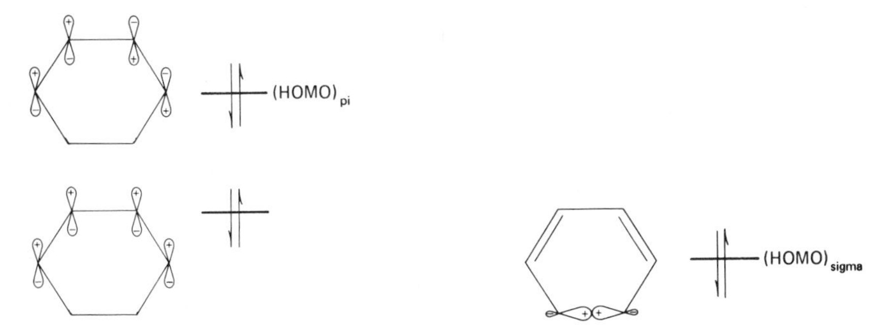

Fig. II.14. *MO's for frontier orbital treatment of the cyclohexadiene ⇌ hexatriene interconversion.*

Once again, his derivation indicates that the most important interactions, energetically, will be $(\text{HOMO})_{\text{sigma-bond}}$–$(\text{LVMO})_{\text{pi-system}}$ and $(\text{HOMO})_{\text{pi-system}}$–$(\text{LVMO})_{\text{sigma-bond}}$. Consider, for example, the thermal opening of a cyclohexadiene to a hexatriene. The MO's of the pi-system and sigma-bond are shown in Fig. II.14. Note very carefully that the pi-system with which we are concerned is that of the reactant, butadiene, and not that of the product, hexatriene. Let us examine the $(\text{LVMO})_{\text{pi}}$–$(\text{HOMO})_{\text{sigma}}$ interaction. Recalling that these orbitals are contained in the same molecule we may, for convenience, incorporate them in the same structure. Let us examine the consequences of conrotatory and disrotatory opening (Fig. II.15). In the conrotatory opening, a bonding interaction of the opening sigma-bond with a terminal pi-orbital lobe is balanced by an antibonding interaction at the other terminus. Thus, to a first approximation, there is no stabilization. However, the disrotatory pathway leads to bonding interactions between the opening sigma-bond and the cyclic pi-system at both termini. Thus, a net stabilization should occur and the disrotatory process is predicted to be the favored one.

PROBLEM II.10

Show that consideration of $(\text{HOMO})_{\text{pi}}$ and $(\text{LVMO})_{\text{sigma}}$ interactions leads to the same prediction regarding ring opening of 1,3-cyclohexadiene.

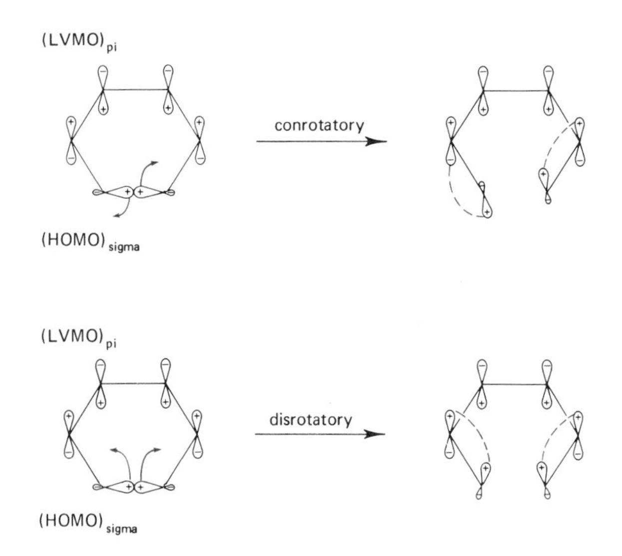

Fig. II.15. *HOMO–LVMO interactions for conrotatory and disrotatory opening of cyclohexadiene.*

PROBLEM II.11

Treat the cyclobutene ⇌ butadiene interconversion by the same technique. What is the predicted mode of thermal conversion?

Fukui showed that the approach outlined above led to the same predictions reached by Woodward and Hoffmann in their first article on concerted reactions. In that paper they argued, with support of extended Hückel calculations, that the symmetry of the highest occupied molecular orbital of the acyclic polyene determines the direction of ring closure.

PROBLEM II.12

Predict the mode of the thermal cyclobutene ⇌ butadiene conversion by consideration of $(\text{HOMO})_{\text{butadiene}}$ symmetry (the relevant MO's may be found in Fig. II.12). Similarly, predict the course of photochemical transformation.

PROBLEM II.13

Predict the course of thermal cyclization of
(a) pentadienyl cation;
(b) pentadienyl anion.
(See Problem II.9 for relevant MO's.)

Woodward and Hoffmann have ascribed the preference of the Diels-Alder reaction for *endo* rather than *exo* cycloaddition to secondary forces arising from interaction of frontier orbitals. Consider the cycloaddition of cyclopentadiene. The orbitals involved in actual bond formation are connected to their bonding partners by black lines in Fig. II.16. The major difference in the two cycloaddition pathways is the presence, in the *endo* addition, of an additional, secondary interaction of the starred orbitals. The interaction is absent in the *exo* addition. The question then arises whether that interaction is favorable or unfavorable. Again, we analyze the HOMO–LVMO interactions (Fig. II.17). As can be seen, the interaction at the starred orbitals is *favorable*. Since it is absent in *exo* attack, *endo* attack should be favored. Again, it should be emphasized that this is a "secondary attractive force." Thus, it may be expected that, in some instances, steric factors may be of greater magnitude than this electronic effect. However, it should also be mentioned that this approach is general and may be applied to other cycloaddition reactions where secondary interactions can occur.

PROBLEM II.14

Assuming electronic factors to be determinative in the $[_{\pi}6_s + {}_{\pi}4_s]$ cycloaddition of 1,3,5-hexatriene to 1,3-butadiene, predict whether *exo* or *endo*

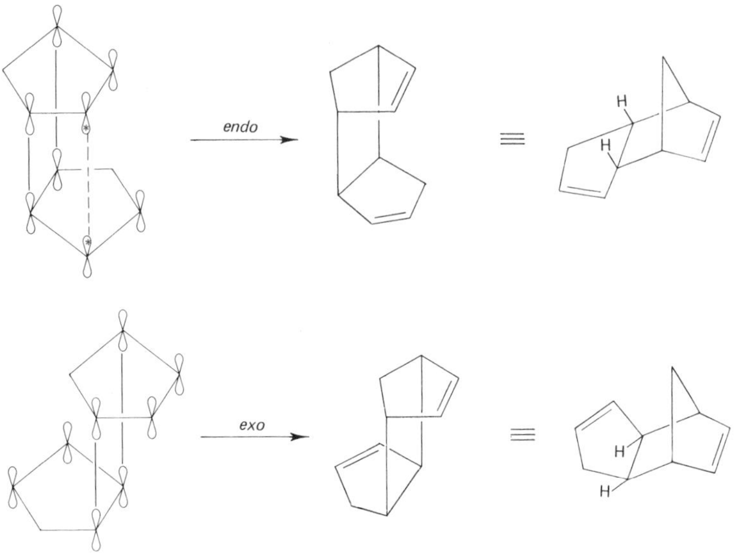

endo

exo

Fig. II.16. *Transition states for exo and endo cycloaddition of cyclopentadiene.*

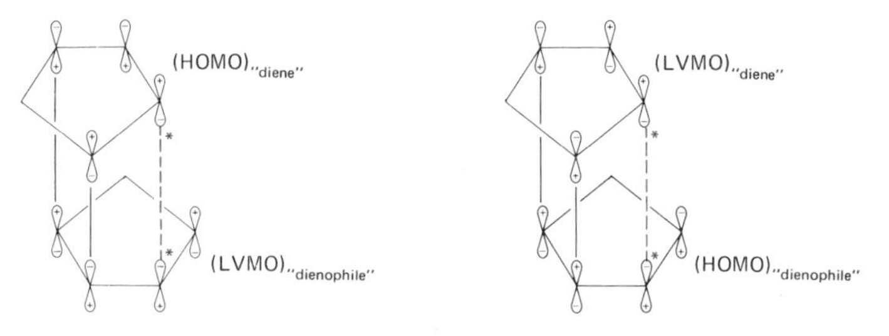

Fig. II.17. *Secondary interactions in endo transition state for cyclopentadiene dimerization.*

cycloaddition should occur. Butadiene MO's can be found in Fig. II.12. The 1,3,5-hexatriene MO's are shown below. (Assume the molecule to lie in the plane of the paper, with the indicated signs corresponding to the sign of the lobe pointing toward the reader.) Consider the secondary orbital interactions at C_2 and C_5 of the hexatriene molecule to be greater than those at C_3 and C_4 (since the overlap at C_2 and C_5 can be seen from inspection of molecular models to be greater than the corresponding overlap at C_3 and C_4).

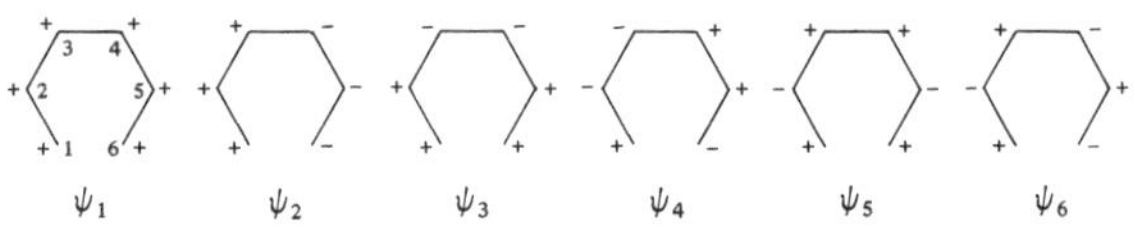

ψ_1 ψ_2 ψ_3 ψ_4 ψ_5 ψ_6

Sigmatropic reactions have been analyzed by Woodward and Hoffmann. While realizing that bonding is maintained throughout the reaction, one imagines the migrating bond to undergo homolytic cleavage and asks whether the migrating atom can pass to the position in question while maintaining bonding interactions. The crucial orbitals examined are the HOMO's of the two species produced by homolytic cleavage. Consider, for example, the suprafacial [1,5] sigmatropic shift of hydrogen [Eq. (II.3)].

(II.3)

Homolytic cleavage of the sigma-bond yields a pentadienyl radical and a hydrogen atom. The HOMO of the pentadienyl radical is then the nonbonding molecular orbital (NBMO) shown (Fig. II.18). Thus, assuming the interaction of the hydrogen

Fig. II.18. *HOMO of pentadienyl radical.*

orbital with the pentadienyl HOMO orbital to be decisive in controlling the course of reaction, it is clear that the hydrogen atom is allowed to shift suprafacially in the [1,5] fashion, but is forbidden to shift in the [1,3] manner (Fig. II.19).

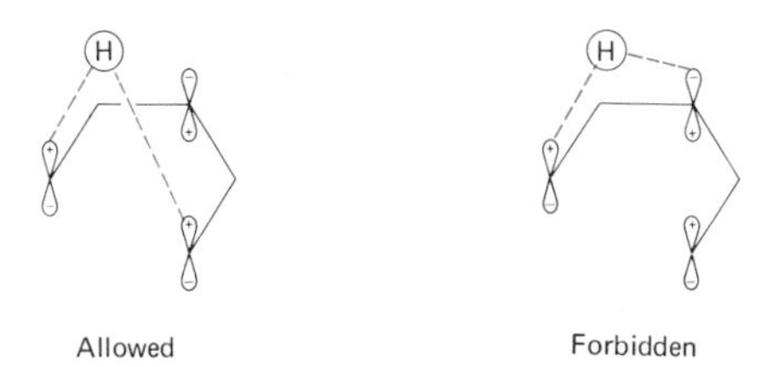

Fig. II.19. *Orbital interactions for* [1,3] *and* [1,5] *hydrogen shifts.*

PROBLEM II.15

Using the approach outlined above, show that a thermal [1,7] antarafacial hydrogen shift should be allowed. (Note: the symmetry of the NBMO of odd-alternant hydrocarbons is always , where "0" indicates the presence of a nodal point.)

PROBLEM II.16

Analyze a thermal [1,3] suprafacial shift of a carbon atom that proceeds with inversion at the migrating atom by the above method. Is it forbidden or allowed?

C. The PMO Method

Dewar has applied perturbational molecular orbital (PMO) theory to concerted reactions. His method involves direct examination of the transition state for a process, and the rules he has derived are notable for their simplicity:

> Thermal electrocyclic (or, more generally, pericyclic) reactions take place via aromatic transition states.
>
> Photochemical electrocyclic (or, more generally, pericyclic) reactions take place through excited forms of antiaromatic transition states.

To apply the rules, however, it is necessary to know how to determine if a particular transition state is aromatic. The aromaticity of a given cyclic, conjugated system is determined by comparison of its pi-energy with that of a corresponding localized polyene. If its pi-energy is less, it is aromatic; if its pi-energy is greater, it is antiaromatic. The pi-energy differences are determined most simply by application of PMO theory. The concept of *union* is essential to its application. Union involves the combination of pi-systems to produce a larger pi-system. Sigma-bonds will be broken and formed, but it is the pi-system and its energy that will concern us. A few examples of union, which is denoted by the symbol $\leftarrow\cup\rightarrow$, are shown in Eqs. (II.4), (II.5), and (II.6).

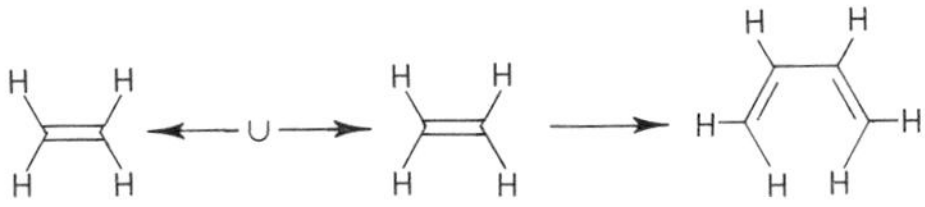

(II.4)

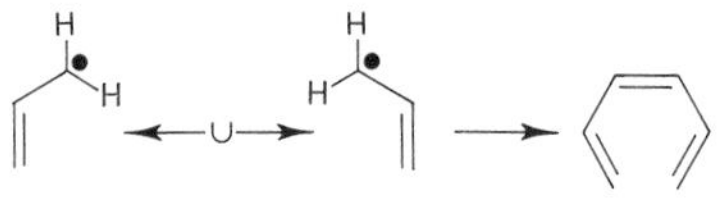

(II.5)

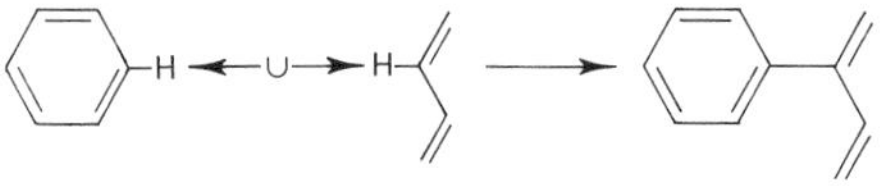

(II.6)

Note that union of two allyl units yields 1,3,5-hexatriene, not biallyl. Dewar has shown that the PMO method enables one to easily calculate the pi-energy change in many such processes. Furthermore, the approach is applicable to the question of aromaticity. For example, union of methyl with pentadienyl at one terminus will yield 1,3,5-hexatriene, whereas union at both termini will yield benzene [Eq. (II.7)].

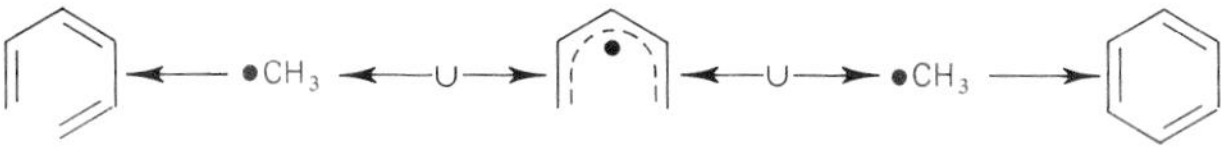

$•CH_3$ ← U → ← U → $•CH_3$ (II.7)

If we are able to calculate the differences in pi-energy resulting from the two processes, we can determine if benzene is aromatic. Fortunately, it is quite easy to calculate the pi-energy of a compound when it can be formed by union of two *odd-alternant* radicals. An odd-alternant system is composed of an odd number of conjugated atoms which can be divided into two sets, a "starred" and an "unstarred" set, in such a way that no two atoms of a given set are directly linked. The "starred" set is chosen as the more numerous. A general property of odd-alternant hydrocarbons (odd AH's) is that the sum of the coefficients of the atoms linked to an unstarred atom is zero in the NBMO. Also, the coefficient at the unstarred atom is zero. The NBMO's of a few representative odd AH's are shown in Fig. II.20.

The numerical value of "a" could be determined by normalization, but a knowledge of the relative signs of the coefficients is usually sufficient and we shall not calculate them. The change in pi-energy upon union of two odd AH's, R and S, is given by Eq. (II.8),

$$\Delta E_\pi = \Sigma 2 a_{0r} b_{0s} \beta_{rs} \qquad \text{(II.8)}$$

where a_{0r} and b_{0s} are the coefficients in the NBMO's of atoms r and s in systems R and S, respectively, and β_{rs} is the resonance integral between atoms r and s.

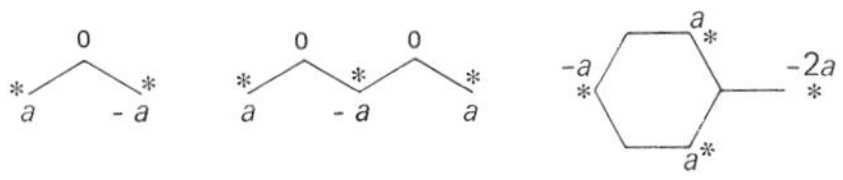

Fig. II.20. *NBMO's of some odd-alternant hydrocarbons.*

If the change in pi-energy is greater for formation of the cyclic structure than for the acyclic analog, it is aromatic. If the two energy changes are the same, the cyclic structure is nonaromatic. If the change in pi-energy is less for the cyclic structure, it is antiaromatic. Let us now return to the question of benzene aromaticity. This is resolved by comparing the ΔE_π's [Eqs. (II.9) and (II.10)].

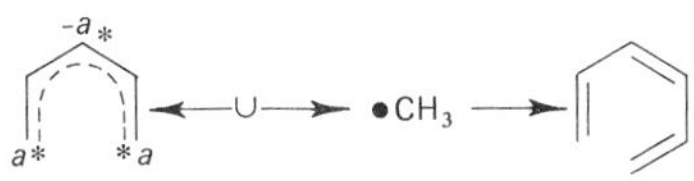

← U → $•CH_3$ → (II.9)

$\Delta E_\pi = 2\beta(1 \cdot a) = 2a\beta$

(the NBMO coefficient of methyl is 1)

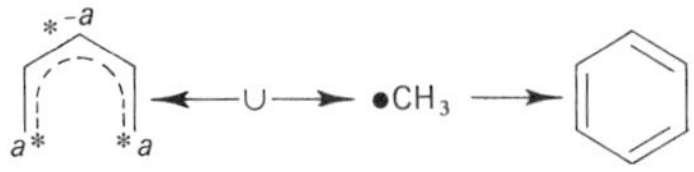

$$\Delta E_\pi = 2\beta(1 \cdot a + 1 \cdot a) = 4a\beta \qquad \text{(II.10)}$$

Since the pi-energy (stabilization) is greater for benzene formation, benzene is aromatic. Cyclobutadiene, on the other hand, is easily shown to be antiaromatic [Eqs. (II.11) and (II.12)].

$$\Delta E_\pi = 2\beta(1 \cdot a) = 2a\beta \qquad \text{(II.11)}$$

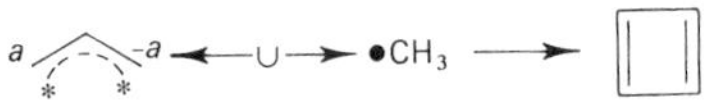

$$\Delta E_\pi = 2\beta(1 \cdot a + 1 \cdot (-a)) = 0 \qquad \text{(II.12)}$$

The acyclic structure is more stable, so cyclobutadiene is antiaromatic.

PROBLEM II.17

Determine whether the following compounds are aromatic, nonaromatic, or antiaromatic:

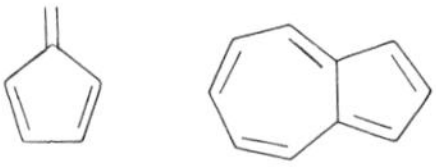

However, an important point remains. We have examined aromaticity in "Hückel" systems, in which the phases of the basis set of AO's are all the same. The results of the PMO treatment indicate that for monocyclic systems Hückel's rule holds: monocyclic, planar conjugated systems with $(4n + 2)$ pi-electrons are aromatic, but those with $4n$ pi-electrons are antiaromatic. However, it can be shown that, if there is an odd number of out-of-phase overlaps, the rules for aromaticity are reversed. For these anti-Hückel, or Möbius, systems monocyclic conjugated structures with $4n$ pi-electrons are aromatic and those with $(4n + 2)$ pi-electrons are antiaromatic! Furthermore, some transition states in concerted reactions are of the anti-Hückel type!* For example, if one examines the transition state for conrotatory and disrotatory cyclobutene ⇌ butadiene interconversion, it is clear that they differ fundamentally (Fig. II.21). The transition state for the disrotatory process is of the Hückel type (all in-phase overlaps), whereas it is

*The existence of Hückel and Möbius transition states has also been clearly recognized by Professor H. E. Zimmerman of the University of Wisconsin. References to his important articles are included at the end of this chapter. His approach differs from that of Dewar primarily by operating within the Hückel method, rather than utilizing PMO theory.

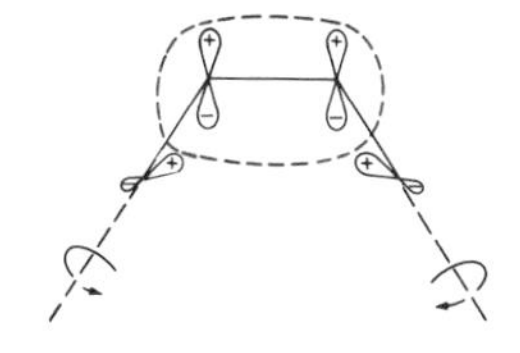

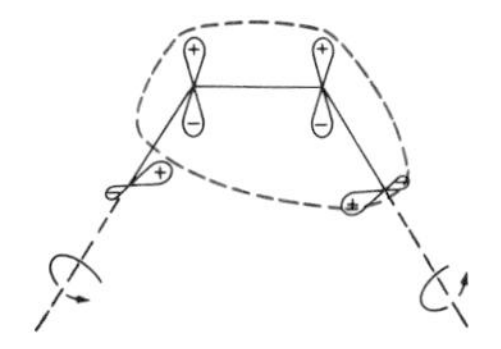

Transition state for disrotatory process

Transition state for conrotatory process

Fig. II.21. *Orbital interactions for disrotatory and conrotatory processes, for the particular basis sets shown.*

impossible to avoid an odd number of out-of-phase overlaps in the conrotatory process, so it is anti-Hückel. The reader may wish to change the phases of individual AO's in the structures cited above (i.e., to change the "basis sets"). The number of out-of-phase overlaps for the Hückel (disrotatory) transition state will always be even, whereas the number of out-of-phase overlaps for the anti-Hückel (conrotatory) transition state will be odd. The crucial question is, then, which transition state is aromatic? This is a $4n$-electron cyclic system that is *isoconjugate* with cyclobutadiene. The system with which a transition state is isoconjugate is determined by an examination of the number of electrons involved and how they are delocalized. In the case at hand, four orbitals defining a ring overlap. Since four electrons are involved, the transition state is isoconjugate with cyclobutadiene. Cyclobutadiene is antiaromatic if of the Hückel type of structure but aromatic if it is anti-Hückel. Thus, the conrotatory process is that which possesses an aromatic transition state and should be favored.

PROBLEM II.18

Analyze the cyclohexadiene ⇌ hexatriene interconversion by the PMO method. Which transition state is aromatic in this instance?

Cycloaddition reactions are easily handled in the same fashion. One simply examines the transition state for a given process and determines whether it is aromatic. The $[_{\pi}4_s + {}_{\pi}2_s]$ Diels-Alder reaction, then, is allowed because the transition state is of the Hückel type and is isoconjugate with benzene, a Hückel aromatic system (Fig. II.22). The reader will again note that changing any

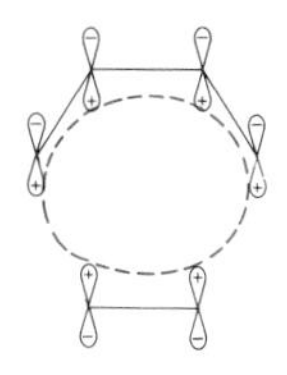

Fig. II.22. *Transition state for* $[_{\pi}4_s + {}_{\pi}2_s]$ *cycloaddition reaction.*

number of phases of AO's will lead to an *even number* of out-of-phase interactions. Similarly, a $[_{\pi}4_s + {}_{\pi}4_s]$ cycloaddition should be forbidden thermally because the transition state is of Hückel type and is isoconjugate with cyclooctatetraene (I), a Hückel antiaromatic system (Fig. II.23). However, the $[_{\pi}4_s + {}_{\pi}4_a]$ thermal cycloaddition should be allowed on electronic grounds since it would proceed via an anti-Hückel transition state (II), which should be aromatic ($4n$ electrons).

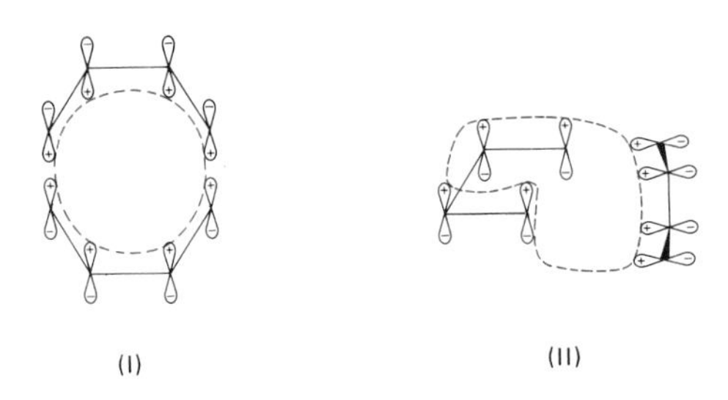

Fig. II.23. *Transition states for* $[_{\pi}4_s + {}_{\pi}4_s]$ *and* $[_{\pi}4_s + {}_{\pi}4_a]$ *cycloaddition reactions.*

Similarly, sigmatropic reactions are easily treated by the PMO approach. Thus, one can examine the transition state for a [1,5] suprafacial hydrogen shift (Fig. II.24). It is clearly of the Hückel type and is isoconjugate with benzene, a Hückel aromatic system. The reaction is therefore allowed. Similarly, antarafacial hydrogen shifts will lead to anti-Hückel transition states. Thus, an antarafacial [1,7] hydrogen shift is allowed (Fig. II.25). As previously stated, the rules are reversed for photochemical reactions.

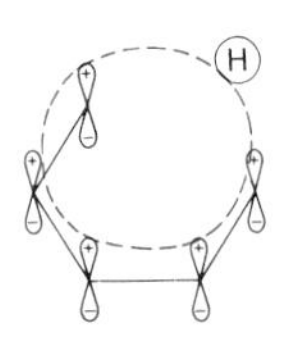

Fig. II.24. *Transition state for a suprafacial* [1,5] *sigmatropic shift.*

PROBLEM II.19

Use the PMO method to analyze [1,3] sigmatropic shifts of a carbon atom which proceed with

(a) retention;

(b) inversion.

Which of the processes is preferred?

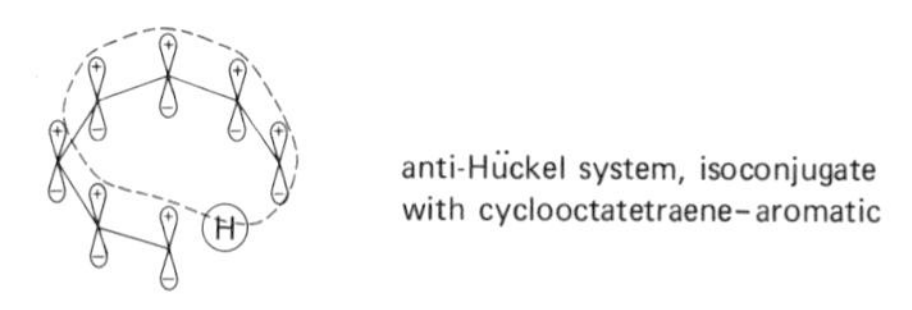

Fig. II.25. *Transition state for antarafacial* [1,7] *sigmatropic shift.*

PROBLEM II.20

The [3,3] sigmatropic rearrangement known as the Cope rearrangement can be imagined to proceed by two different stereochemical pathways:

(1)

or

(2)

Examine the transition states in each case to determine which process should be favored.

Suggested Supplemental Readings

The books and articles listed here may be consulted by the interested reader to supplement topics discussed in Chapter II.

J. D. Roberts, "Notes on Molecular Orbital Calculations." Benjamin, New York, 1962. This paperback provides an introduction to the Hückel method and will be helpful to the student who is unfamiliar with molecular orbital theory.

R. B. Woodward and R. Hoffmann, "The Conservation of Orbital Symmetry." Verlag Chemie, Weinheim, and Academic Press, New York, 1970. This is the most thorough treatment of pericyclic reactions available. Both theoretical background and copious examples are included. This excellent book will be especially valuable to the reader wishing to understand more fully the basic bonding processes involved in pericyclic reactions and, especially, the construction of correlation diagrams.

M. J. S. Dewar, "The Molecular Orbital Theory of Organic Chemistry." McGraw-Hill, New York, 1969. The theoretical basis for the PMO method is developed in this text, and numerous examples of its application both to pericyclic and to other reactions are included (see especially Chapters 6 and 8). See also M. J. S. Dewar, *Angew. Chem. Int. Ed. Engl.* (1971) (in press).

K. Fukui and H. Fujimoto, *in* "Mechanisms of Molecular Migrations," (B. S. Thyagarajan, ed.), Vol. 2, pp. 118–186. Wiley, New York, 1969. The authors treat a variety of pericyclic reactions, primarily from the standpoint of frontier orbital theory. For a more recent treatment, see K. Fukui, *Accounts Chem. Res.* **4**, 57 (1971).

H. E. Zimmerman, *J. Amer. Chem. Soc.* **88**, 1565 and 1566 (1966). Professor Zimmerman focuses attention upon Hückel and Möbius systems, and shows how simple circle mnemonics can be used to analyze electrocyclic reactions. A more recent article by Professor Zimmerman can be found in the following reference: H. E. Zimmerman, *Accounts Chem. Res.* **4**, 272 (1971).

H. C. Longuet-Higgins and E. W. Abrahamson, *J. Amer. Chem. Soc.* **87**, 2045 (1965). These authors use state correlation diagrams to analyze electrocyclic reactions.

R. G. Pearson, *Accounts Chem. Res.* **4**, 152 (1971). Another approach is presented that entails the application of MO theory to pericyclic and other reactions.

Part II

PROBLEMS

Chapter III

CYCLOADDITION REACTIONS

PROBLEM III.1

(a) Suggest a mechanism for the following reaction:

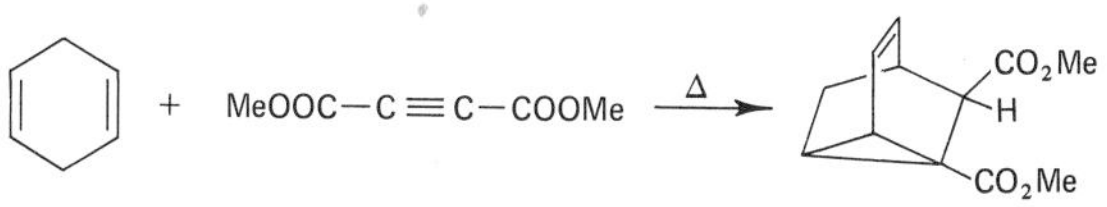

+ MeOOC—C≡C—COOMe $\xrightarrow{\Delta}$ [product: CO_2Me, H, CO_2Me]

(b) Norbornadiene is known to undergo $[_{\pi}2_s + _{\pi}2_s + _{\pi}2_s]$ thermal cycloaddition with maleic anhydride.* Suggest a structure for the adduct.

(c) With your answers to parts (a) and (b) of this question in mind, suggest possible structures for the anticipated products of the reaction of isotetralin with dimethyl acetylenedicarboxylate.

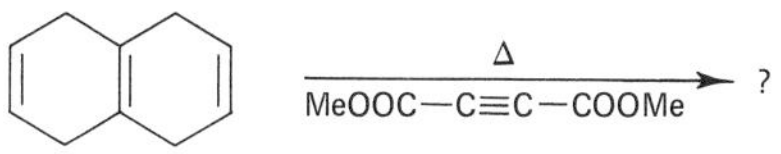

$\xrightarrow[\text{MeOOC—C≡C—COOMe}]{\Delta}$?

PROBLEM III.2

The photochemical reaction of acetylenedicarboxylic acid with 1,4-cyclohexadiene has been found to proceed in the following manner:

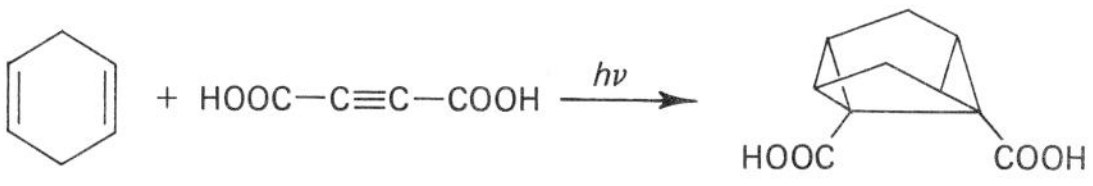

+ HOOC—C≡C—COOH $\xrightarrow{h\nu}$ HOOC COOH

(I)

(a) Classify this reaction as an $[m + n + o + \cdots]$ cycloaddition process. (I.e., what are the values of $m, n, o, \cdots$?)

(b) Account mechanistically for the formation of I.

*R. C. Cookson, J. Dance, and J. Hudec, *J. Chem. Soc.* p. 5416 (1964).

PROBLEM III.3

Predict the products (show stereochemistry!) of each of the reactions indicated below. Assume in all cases concerted reactions which proceed to give the symmetry-allowed product. Where more than one product can be formed by symmetry-allowed processes, draw all possible products.

(a) 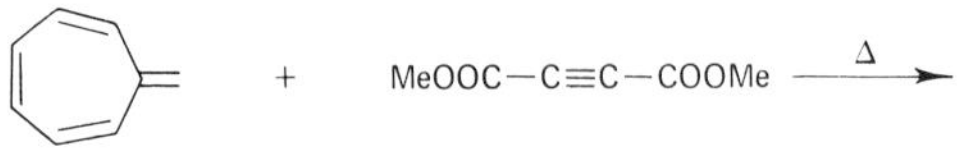 + MeOOC—C≡C—COOMe $\xrightarrow{\Delta}$ 1:1 adduct

(b) 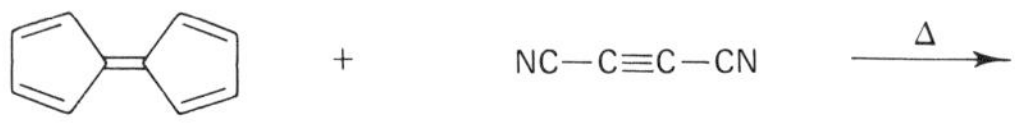 + NC—C≡C—CN $\xrightarrow{\Delta}$ 1:2 adduct

(c) 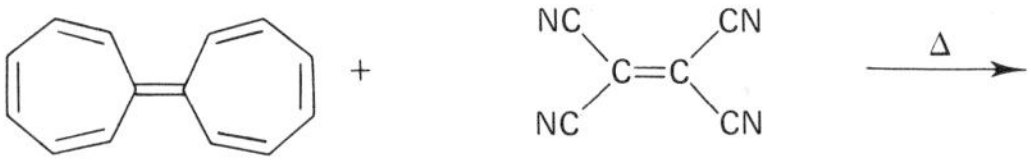 + (NC)$_2$C=C(CN)$_2$ $\xrightarrow{\Delta}$ 1:1 adduct

(d) 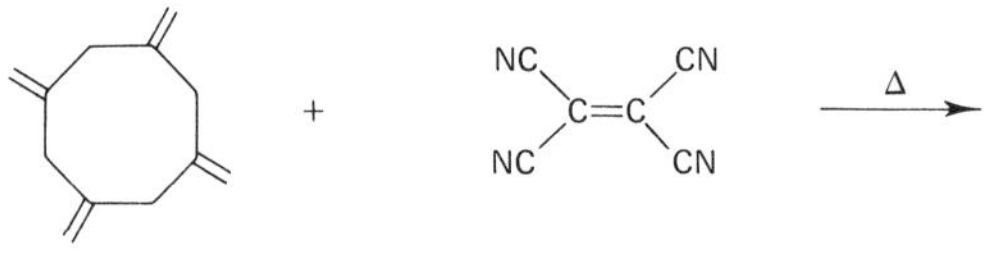 + (NC)$_2$C=C(CN)$_2$ $\xrightarrow{\Delta}$ 1:1 adduct

(e) 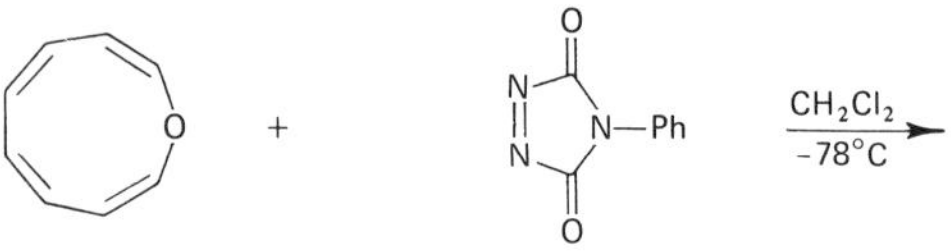 O + N—Ph $\xrightarrow[-78°C]{CH_2Cl_2}$ 1:1 adduct containing two conjugated carbon-carbon double bonds

(f) 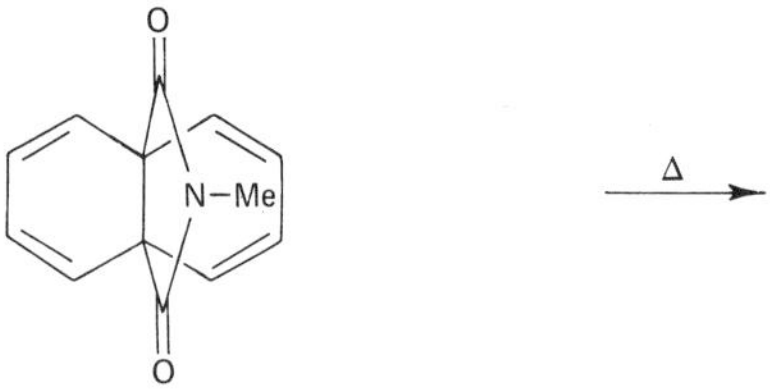 N—Me $\xrightarrow{\Delta}$ dimer containing four carbon-carbon double bonds

PROBLEM III.4

Consider the thermal reaction of *cis*-bicyclo[6.1.0]nona-2,4,6-triene (I) with maleic anhydride:

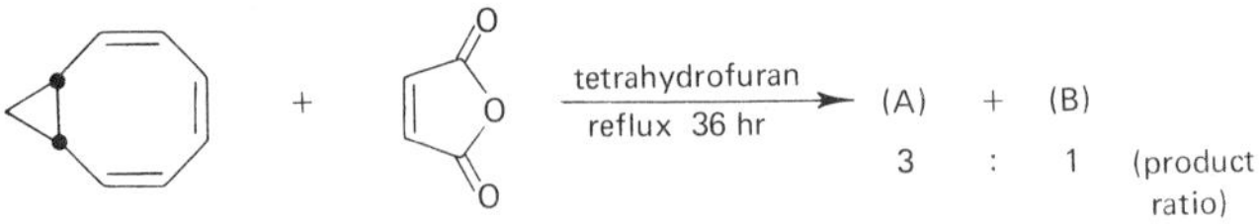

(I) + O $\xrightarrow[\text{reflux 36 hr}]{\text{tetrahydrofuran}}$ (A) + (B)

3 : 1 (product ratio)

The following information is pertinent:

(1) Both compounds A and B are 1:1 adducts.

(2) Compound A contains one carbon–carbon double bond.

(3) Compound B contains two carbon–carbon double bonds.

Identify compounds A and B and suggest mechanisms for their formation from I.

PROBLEM III.5

Consider the processes (1) through (4):

(1) 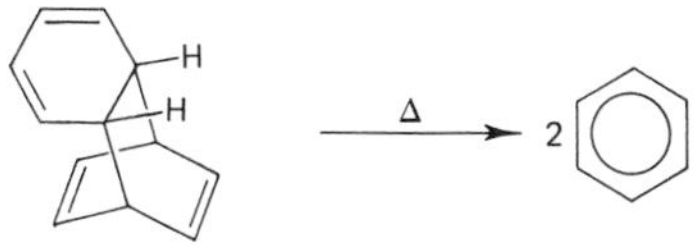 $\xrightarrow{\Delta}$ 2

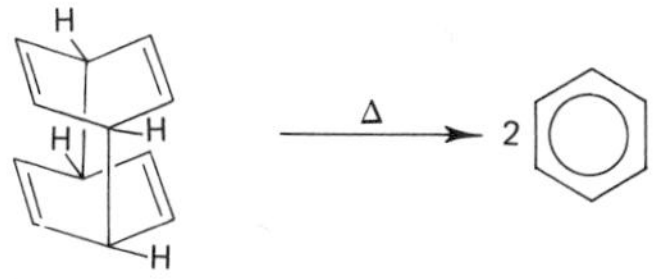

(2) $\xrightarrow{\Delta}$ 2

(3) 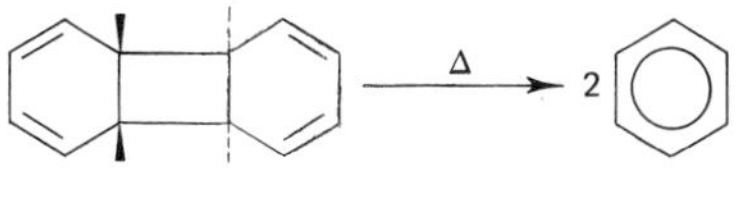 $\xrightarrow{\Delta}$ 2

(4) 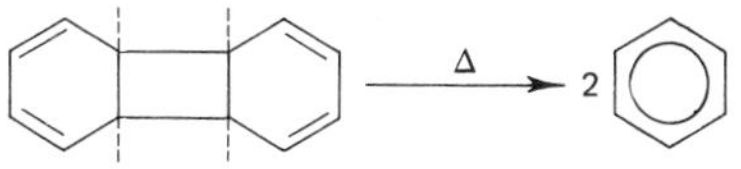 $\xrightarrow{\Delta}$ 2

Classify each of these reactions as an $[m + n]$ cycloreversion. Which reaction would be expected to be most facile? Explain.

PROBLEM III.6

(a) The following reaction sequence has been observed:

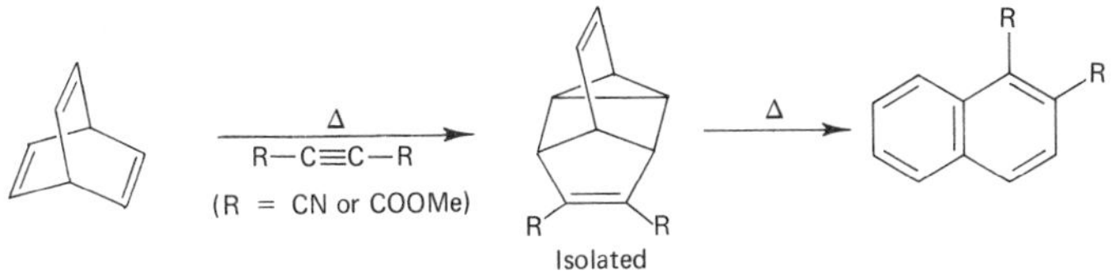

$\xrightarrow[\text{R—C}\equiv\text{C—R}]{\Delta}$ (R = CN or COOMe) $\xrightarrow{\Delta}$

Isolated

The second step of the reaction is postulated to proceed via an intermediate 1,2-disubstituted *cis*-9,10-dihydronaphthalene which then aromatizes by intramolecular hydrogen transfer.

Suggest a detailed mechanism for each step of the above reaction sequence.

(b) Dr. T. L. Cairns and co-workers at E. I. duPont de Nemours and Company have observed an interesting cycloaddition reaction of substituted acetylenes to benzene:

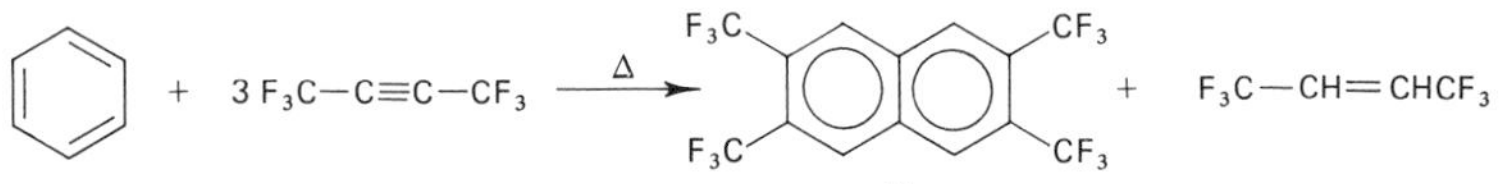

+ 3 F₃C—C≡C—CF₃
Δ
F₃C
CF₃
F₃C
CF₃
+ F₃C—CH=CHCF₃

(I)

This same group of investigators has shown that bis(trifluoromethyl)acetylene adds in Diels-Alder fashion to durene to give a highly substituted barrelene, II:

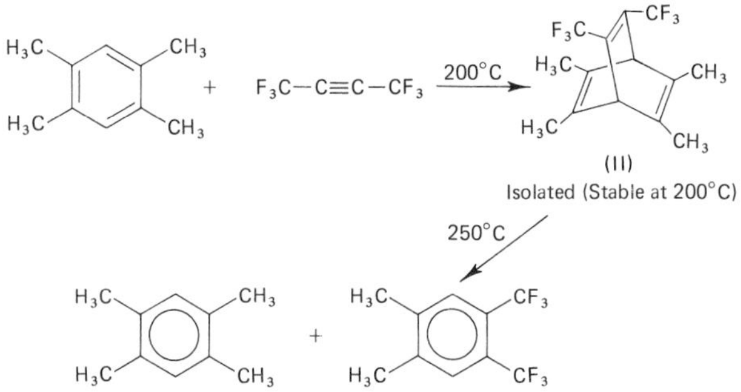

H₃C
CH₃
H₃C
CH₃
+ F₃C—C≡C—CF₃
200°C
F₃C
CF₃
H₃C
CH₃
H₃C
CH₃
(II)
Isolated (Stable at 200°C)
250°C
H₃C
CH₃
H₃C
CH₃
+
H₃C
CF₃
H₃C
CF₃

(Major volatile products)

Suggest a mechanism for the indicated preparation of compound I proceeding via a barrelene-type intermediate. On the basis of orbital symmetry considerations, comment on the "symmetry-allowedness" of each step of your mechanism. Predict the stereochemistry of the bis(trifluoromethyl)-ethylene (*cis* or *trans*?) which is formed along with I in the reaction.

PROBLEM III.7

Thermolysis of cycloheptatriene affords a dimeric product containing two carbon–carbon double bonds. Suggest a structure for this dimer and suggest a mechanism for its formation.

PROBLEM III.8

Sydnones are a class of mesoionic compounds which display aromatic properties. They are, however, capable of undergoing cycloaddition reactions. A particularly interesting example is the reaction of 3-phenylsydnone (I) with *cis,cis*-cycloocta-1, 5-diene to produce the novel system, II:

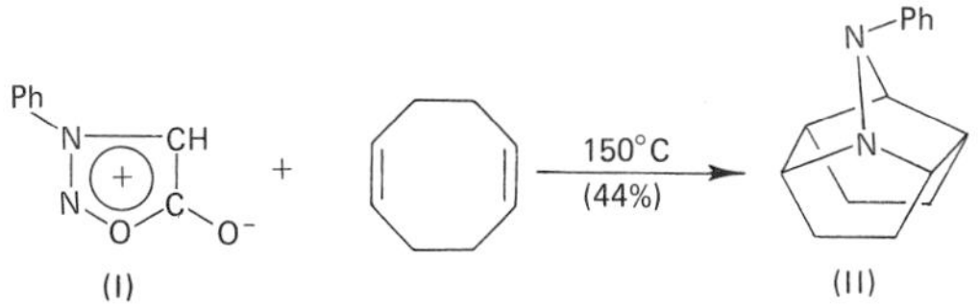

Ph N CH + N C O O⁻ (I) + 150°C (44%) N–Ph N (II)

Suggest a mechanism for this transformation via a symmetry-allowed pathway.

PROBLEM III.9

The following cycloreversion reactions have recently been studied:

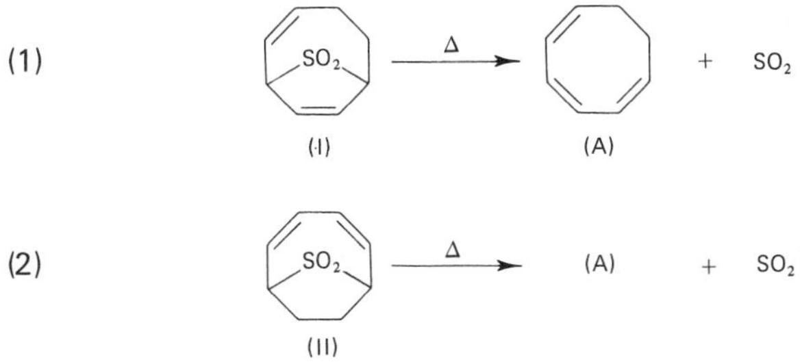

(1) SO_2 (I) $\xrightarrow{\Delta}$ (A) + SO_2

(2) SO_2 (II) $\xrightarrow{\Delta}$ (A) + SO_2

Thermochemical Data:

	Process (1)	Process (2)
ΔH^* (kcal/mole)	31.2 ± 0.9	30.0 ± 1.5
ΔS^* (eu)	+ 7. ± 3.	− 18. ± 5.
ΔF^* (kcal/mole)	28.5 ± 0.9	39.5 ± 1.5

Suggest mechanisms to account for the observed $\Delta\Delta F^* =$ ca. 10 kcal/mole for these two cycloreversion reactions.

PROBLEM III.10

Cyclobutadiene, produced *in situ* by ceric ion oxidation of cyclobutadieneiron tricarbonyl, reacts with isopyrazole to produce compound A, $C_{21}H_{20}N_2$. Direct photolysis of A affords compound B, $C_{21}H_{20}$ plus nitrogen. Both A and B can be thermally converted to C, $C_{21}H_{20}$. Compound C contains no *olefinic* carbon–carbon double bonds.

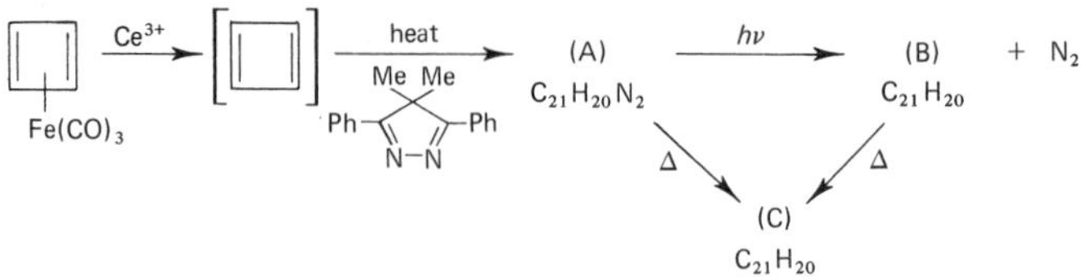

(A) $C_{21}H_{20}N_2$ → (B) $C_{21}H_{20}$ + N_2; (C) $C_{21}H_{20}$

Suggest structures for compounds A, B, and C, and provide mechanisms to account for their formation.

PROBLEM III.11

The following cycloaddition reaction has been observed:

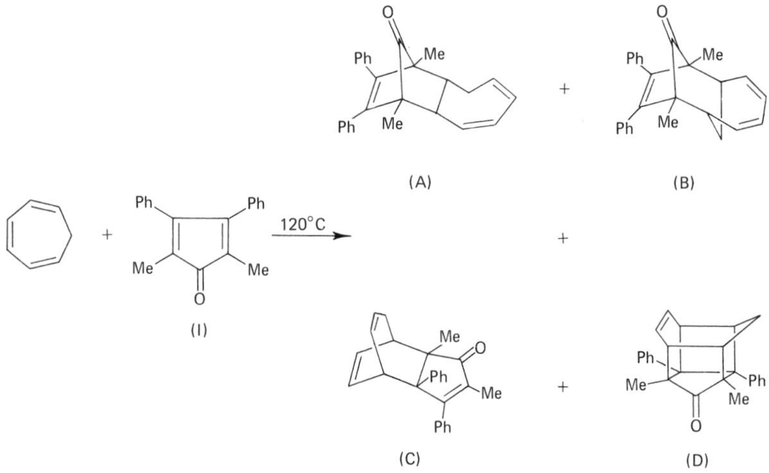

Some pertinent observations are as follows:

(a) The initial cycloaddition affords only A and B. Products C and D are formed subsequently, and at the expense of A and B, respectively.

(b) In the reaction of I with *excess* cycloheptatriene at 170°C, two new products are observed (E and F).

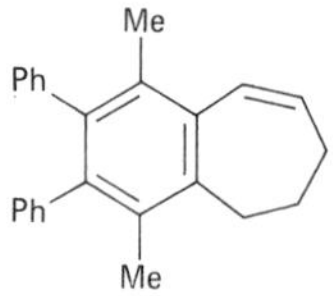

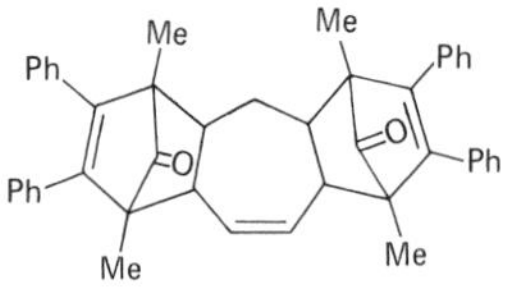

Suggest mechanisms for the formation of each of the adducts A through F.

PROBLEM III.12

The following reaction has been observed:

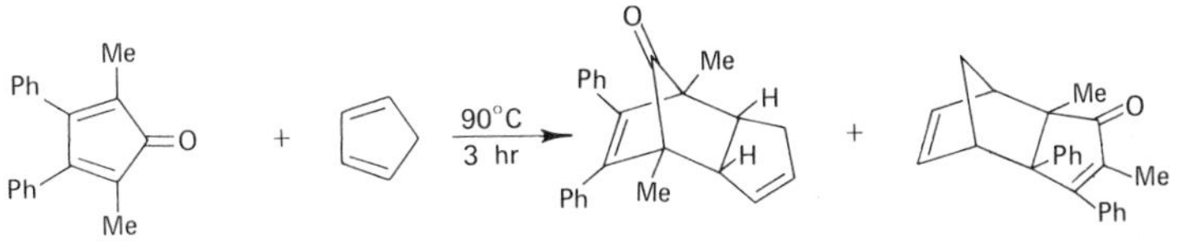

(I) + cyclopentadiene —(90°C, 3 hr)→ (A) 90% + (B) 10%

When compound A is heated at 105° C, an equilibrium is established between A and B; at equilibrium, the ratio of A : B = 1 : 1. Under these conditions, reversal of the Diels-Alder adduct A to I + cyclopentadiene does *not* occur. Compound A is formed essentially stereospecifically *endo*. The reaction of I with cyclopentene, however, is essentially *non*stereospecific.

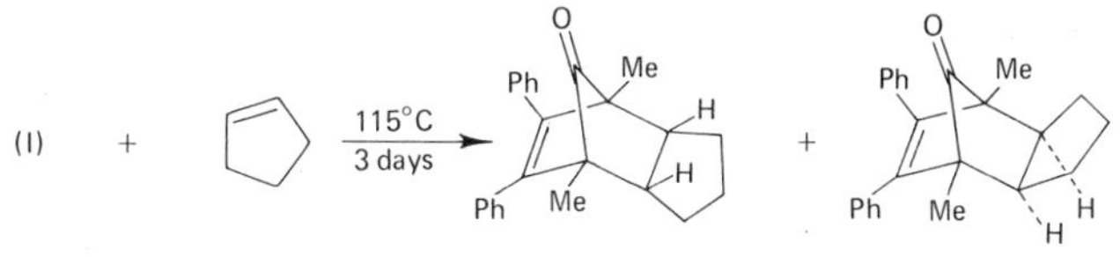

(I) + cyclopentene —(115°C, 3 days)→ (C) + (D)

56 : 44 (product ratio)

(a) Comment on the observed stereospecificity of the reaction of I with cyclopentadiene, as compared with the essentially nonstereospecific nature of the corresponding reaction with cyclopentene.

(b) Suggest a mechanism for the intramolecular conversion of A to B in the equilibration experiment carried out by heating A to 105°C.

PROBLEM III.13

Treatment of indene with a fourfold excess of 2-bromofluorobenzene and magnesium in tetrahydrofuran affords after hydrolysis a new compound, A, $C_{15}H_{10}$. If the final hydrolysis step is carried out using D_2O instead of H_2O, the product A has the molecular formula $C_{15}H_9D$. Suggest a structure for compound A, and suggest a mechanism for its formation which is consistent with the fact that deuterium incorporation occurs under the conditions described above. (What position in the molecule does the deuterium thus incorporated occupy?)

Chapter IV

THE STEREOCHEMISTRY OF SIGMATROPIC REACTIONS

Extensive studies of the stereochemistry of concerted, sigmatropic carbon–carbon bond rearrangements of orders [1,3] and [1,5] have been carried out by Professor J. A. Berson at Yale University and by Professor H. E. Zimmerman at the University of Wisconsin. In this chapter, we shall consider their work in this area to exemplify the constraints imposed on the stereochemical outcome of concerted sigmatropic rearrangements in cases where orbital symmetry is conserved.

PROBLEM IV.1

By way of orientation, the reader should review the sections in Chapter I which deal with the applications of orbital symmetry relationships to sigmatropic rearrangements.

(a) Consider first *suprafacial* sigmatropic rearrangements of order [1,3]:

R [1,3] R

What is the stereochemical fate of the migrating group, R, if this reaction is to proceed thermally in accordance with the Woodward-Hoffmann rules? (I.e., does R retain its configuration or does it suffer inversion in the thermally allowed suprafacial [1,3] sigmatropic rearrangement?)

(b) Repeat your analysis of part (a), above, for a thermal, suprafacial sigmatropic rearrangement of order [1,5].

PROBLEM IV.2

(a) Suggest a mechanism for the following transformation:

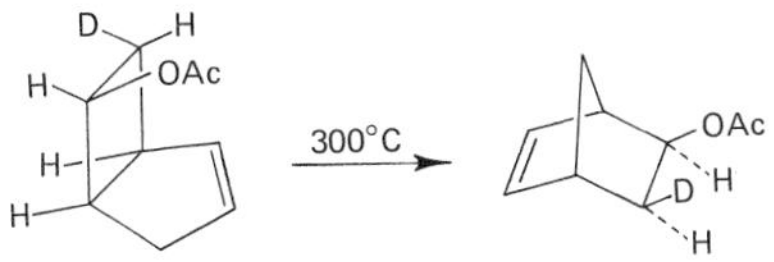

300°C

(b) Consider the following thermal [1,3] sigmatropic rearrangements:

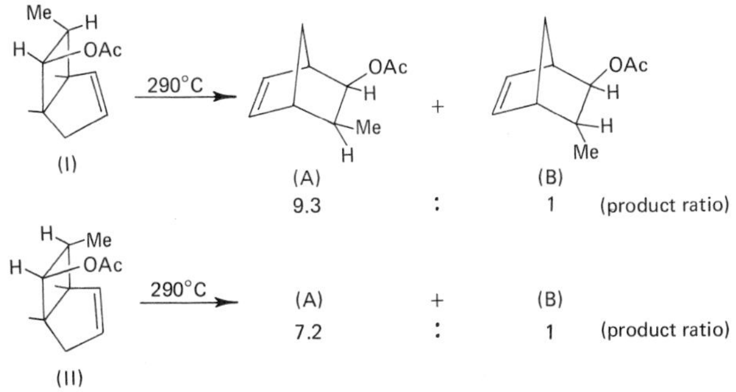

(I) —290°C→ (A) + (B)
9.3 : 1 (product ratio)

(II) —290°C→ (A) + (B)
7.2 : 1 (product ratio)

The rearrangement of *endo,exo*-6-acetoxy-7-methylbicyclo[3.2.0]hept-2-ene (I) is seen to proceed with predominant inversion, whereas rearrangement of the corresponding *endo, endo* isomer proceeds with predominant *retention* of configuration of the migrating center. Furthermore, *exo* ⇌ *endo* epimerization of the methyl group does not compete with rearrangement during the pyrolysis of I, but this process is found to occur 60% as fast as the rearrangement of II to A and B.

Offer a detailed explanation to account for these observations.

PROBLEM IV.3

Cycloheptatrienes are known to undergo thermal [1,5] sigmatropic skeletal rearrangements via norcaradiene intermediates:

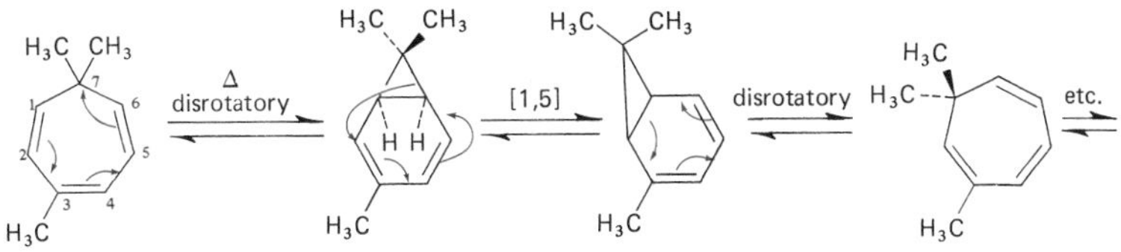

Δ disrotatory ⇌ [1,5] ⇌ disrotatory ⇌ etc.

In principle, one could directly test the predictions of Woodward-Hoffmann theory regarding the stereochemistry of the migrating group, C_7, in this rearrangement sequence. Berson and Willcott* have suggested a method utilizing an optically active tropilidene.

Starting with the optically active cycloheptatriene, I, carry out one complete circuit of interconverting (via [1,5] sigmatropic shifts) norcaradiene intermediates. This should be done for the cases of retention of configuration at C_7 and of inversion at C_7, assuming a suprafacial migration in both cases.

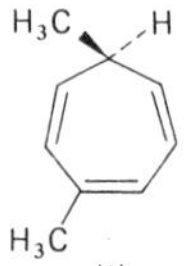

H_3C H
H_3C
(I)

If the rearrangement were performed thermally, show how you could decide on the basis of the stereochemical outcome of the above norcaradiene interconversions whether the rearrangement of I occurred via the symmetry-allowed path or via the symmetry-"forbidden" (photochemically allowed) path.

PROBLEM IV.4

An alternative to the direct [1,5] sigmatropic rearrangement mechanism discussed in Problem IV.3 for the interconversion of the norcaradiene intermediates would be two successive [1,3] sigmatropic shifts; using compound I (Problem IV.3) as substrate, we can formulate this alternative mechanism as follows:

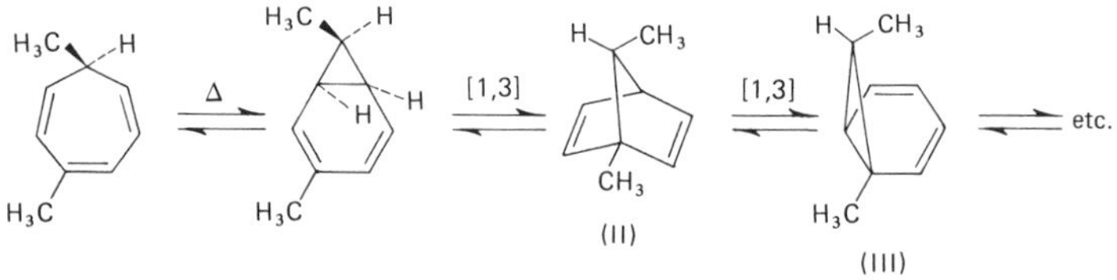

H_3C H H_3C Δ H_3C H H H H_3C [1,3] H CH_3 CH_3 (II) [1,3] H CH_3 H_3C (III) etc.

(The diagram illustrates two successive [1,3] sigmatropic migrations with retention of configuration at C_7. The same configuration of the norcaradiene III would be obtained if inversion at C_7 were incurred in both of the steps.) However, this alternative mechanism has been excluded; the substituted norbornadiene (II) has been shown *not* to be an important intermediate.†

*J. A. Berson and M. R. Willcott, III, *Rec. Chem. Progr.* **27**, 139 (1966).
†M. R. Willcott, III, and C. J. Boriack, *J. Amer. Chem. Soc.* **90**, 3287 (1968).

Zimmerman has suggested a third mechanism for this reaction. Termed the "slither" mechanism, it can be simply formulated as two sequential 1,2-shifts. For the cycloheptatriene system, this amounts to interconversion of the norcaradiene intermediates by the following sequence:

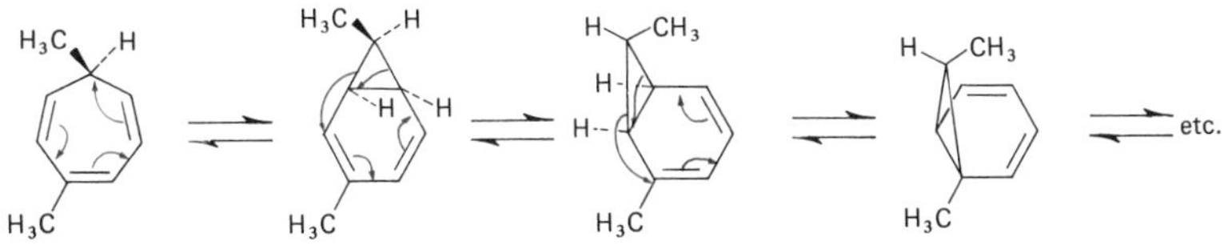

Demonstrate whether or not optical purity would be maintained in the rearrangement of I (Problem IV.3) via the slither mechanism.

PROBLEM IV.5

Zimmerman and Hill and co-workers have studied the stereochemistry of the rearrangement shown below:

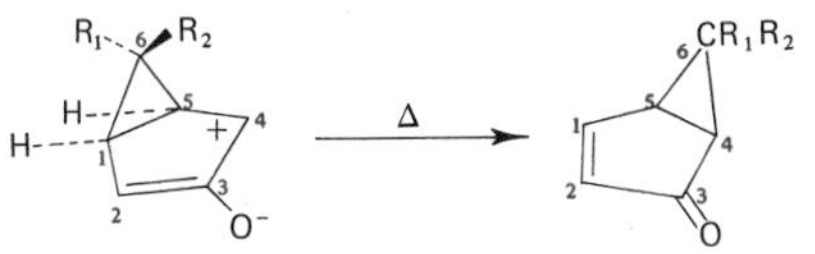

(a) For the indicated rearrangement, determine the stereochemistry at C_6 if
    (1) rearrangement occurs by the slither mechanism;
    (2) rearrangement occurs by a suprafacial [1,4] sigmatropic shift with retention of the migrating atom;
    (3) rearrangement occurs by a suprafacial [1,4] sigmatropic shift with inversion at the migrating atom.
(b) Analyze the reaction theoretically and predict the preferred course of thermal rearrangement.
(c) Rationalize each of the following transformations. Are the results in accord with the prediction you made in part (b)?

(1) KO-*t*-Bu, *t*-BuOH, 42°C (Ar = *p*-bromophenyl)

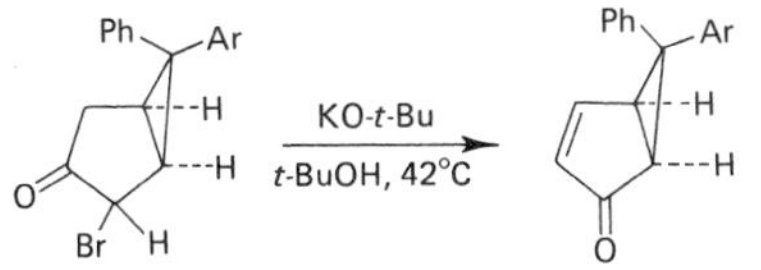

(2) KO-*t*-Bu, *t*-BuOH, 42°C (Ar = *p*-bromophenyl)

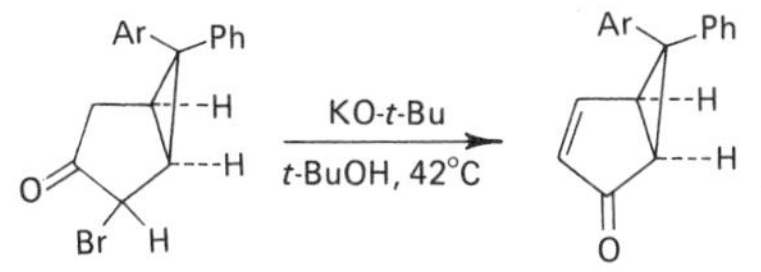

PROBLEM IV.6

The [1,3] sigmatropic rearrangement of optically active Feist's ester (below) is especially interesting. The reaction has been shown to proceed with a high degree of stereospecificity.

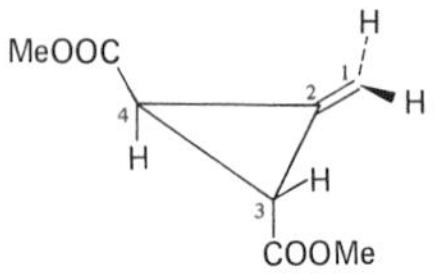

Feist's ester

The thermal [1,3] sigmatropic rearrangement carried out on optically active Feist's ester possessing the configuration shown above proceeds with cleavage of the cyclopropane C_3—C_4 sigma-bond and has been found to yield two optically active products.

(a) What are the symmetry-allowed processes in this system?

(b) Four products can result in this rearrangement. Show how each could arise via a symmetry-allowed pathway.

(c) In fact, only two of the four products are actually observed. Analyze your rearrangements to determine which of the "symmetry-allowed" processes should predominate, and thereby predict structures for the products.

PROBLEM IV.7

The Cope and Claisen rearrangements are both [3,3] sigmatropic shifts. A huge number of examples of these rearrangements appear in the literature. One of these is particularly worthy of note, as will be seen from a detailed examination of the following problem.

Consider the [3,3] sigmatropic rearrangement shown below:

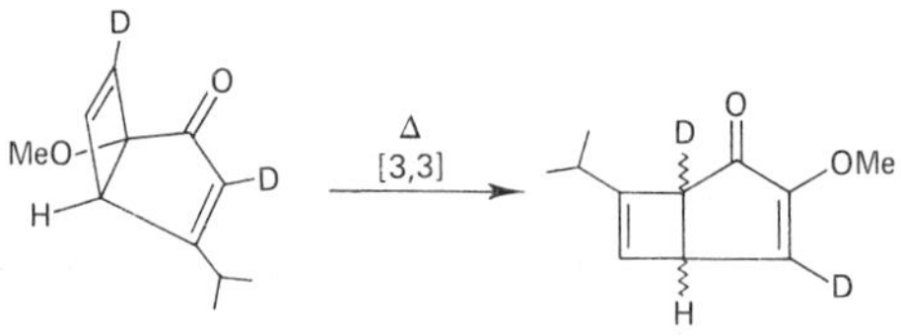

(a) Suggest a mechanism for this reaction. Your mechanism should clearly show the stereochemistry (*cis* or *trans*) of the ring junction in the product.

(b) In terms of a $[_{\sigma}2 + _{\pi}2 + _{\pi}2]$ process, is your mechanism *supra–supra*, *supra–antara*, or *antara–antara* with respect to the pi-bonds?

(c) Doering and Roth* have shown that the preferred transition state in the Cope rearrangement is "chairlike," as shown:

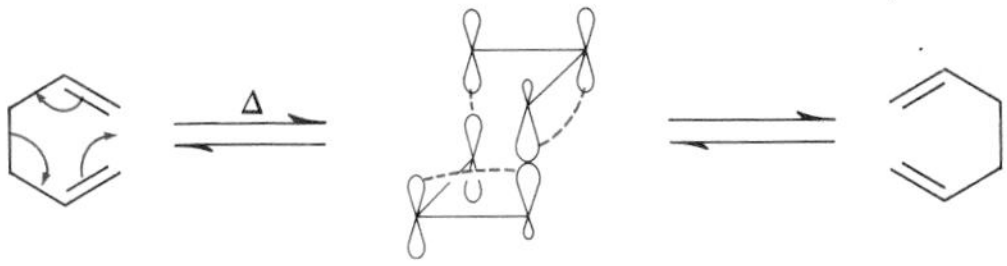

"Chairlike" transition state

This is a $[_{\sigma}2_s + _{\pi}2_s + _{\pi}2_s]$ process, and as such is *supra–supra* with respect to the pi-bonds. In the light of this information, comment on your answer to part (b) of this question. Why is the present example "particularly worthy of note?"

PROBLEM IV.8

The intramolecular, thermal rearrangement of (−)-1-methyl-3-*t*-butylindene (I) has been studied. Optically pure I, when heated to 140°C and partially isomerized, affords compound II, which is 99% racemic.

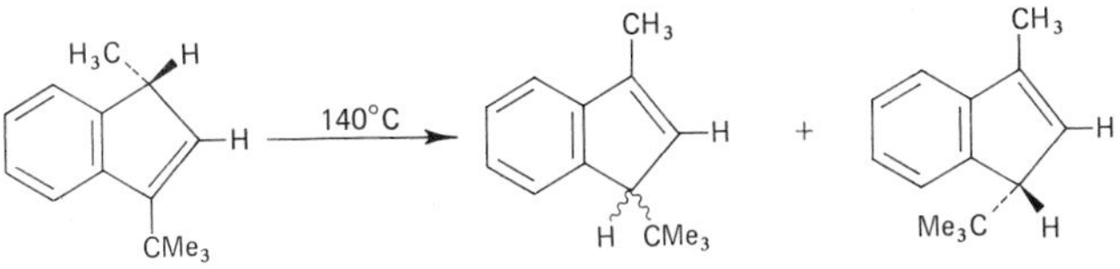

(-)-(I) Optically active — 140°C → (II) 99% racemic + (+)-(II) 1%

The remaining 1% possesses the configuration (+)-II, indicating an overall suprafacial hydrogen shift for this minor reaction pathway.

(a) Suggest a mechanism for the major reaction pathway, (−)-I ⟶ racemic II. Your mechanism must account for the observed loss of optical activity in proceeding from I ⟶ II.

(b) Several pathways can be suggested to explain the formation of (+)-II in the minor reaction. Two of these are listed below:

(1) Direct, concerted thermal [1,3] sigmatropic hydrogen shift

(2) A series of thermal [1,5] sigmatropic shifts along the bridgehead carbon atoms

Draw flow diagrams to pictorially represent each of the above mechanisms, (1) and (2). Analyze them with regard to symmetry-allowedness. Which of the

*W. von E. Doering and W. R. Roth, *Tetrahedron* **18**, 67 (1962).

two mechanisms appears preferable for explaining the course of the minor reaction?

PROBLEM IV.9

The photochemical transformation of bicyclo[3.2.0]hept-3-en-2-one (I) to 7-ketonorbornene (II) has been studied.

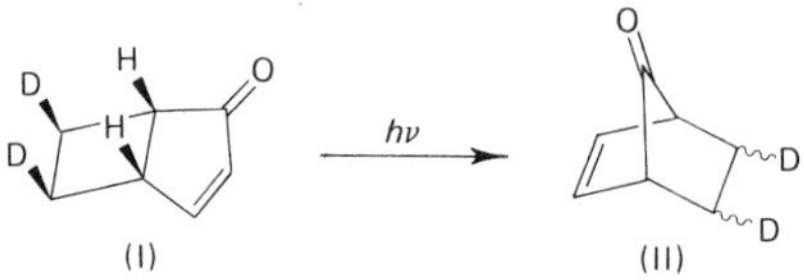

(I) —$h\nu$→ (II)

(a) Predict the stereochemistry of the C—D bonds in the product, II, assuming that the reaction is concerted and proceeds in accordance with the principles of conservation of orbital symmetry. Does the migrating carbon terminus suffer inversion or retention via your mechanism, or would a nonstereospecific migration be anticipated? Explain.

(b) Analysis of the NMR spectrum of II reveals that the rearrangement proceeds with complete loss of stereochemistry at the migrating terminus. Suggest a mechanism which is compatible with this observation [if one different from that which you suggested in part (a) is required].

Chapter V

MOLECULES WITH FLUCTIONAL STRUCTURES

Professor W. von E. Doering at Harvard University has studied a number of compounds which can undergo one or more degenerate [3,3] sigmatropic rearrangements such that each of the rearranged products is identical in structure with the original starting material. An example is homotropilidene:

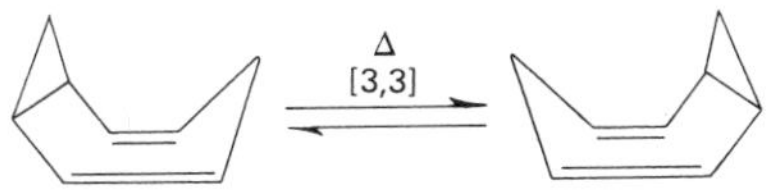

Δ [3,3]

(1a) (1b)

Evidence that the rearrangement Ia ⇌ Ib can occur rapidly is provided by the NMR spectrum of homotropilidene, the appearance of which is found to be strongly temperature-dependent. [For a detailed discussion of the effect of temperature on the NMR spectrum of homotropilidene, see W. von E. Doering and W. R. Roth, *Angew. Chem. Int. Ed. Engl.* **2**, 115 (1963).]

From an analysis of the NMR spectrum, it has been estimated that homotropilidene undergoes the [3,3] sigmatropic rearrangement Ia ⇌ Ib with a frequency $\nu =$ ca. 1 sec^{-1} at $-50°C$ and $\nu =$ ca. 10^3 sec^{-1} at 180°C. Thus, homotropilidene has no single, unique structure, but is best described in terms of a "time average" structure whose extremes are defined by the tautomers Ia and Ib. The term "fluctional" is used to describe the structure of compounds of this type, implying that the actual positions of the atoms in the molecule fluctuate (or oscillate) statistically between their extreme positions as defined by Ia and Ib.

Consideration of models suggests that the [3,3] sigmatropic rearrangement occurs through the cisoid conformation, Ia or Ib.

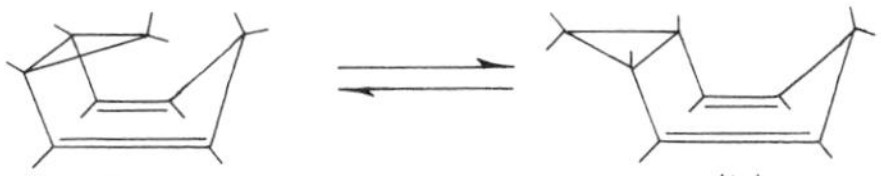

(Ia)
Cisoid

(Ic)
Transoid

However, it is the transoid form (Ic) which predominates in the equilibrium Ia ⇌ Ic. It follows that exclusion of this equilibrium (i.e., constriction of homotropilidene into its cisoid conformation) would necessarily promote the [3,3] sigmatropic rearrangement Ia ⇌ Ib. This can be done by the introduction of a methylene bridge (II), of an ethylene bridge ("bullvalene," III), or by a direct 1,5 sigma-bond ("semibullvalene," IV).

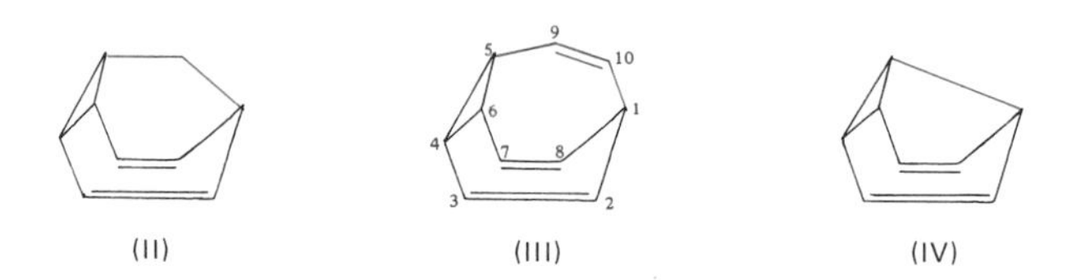

The remarkable compound bullvalene, III, has a threefold axis of symmetry; all three double bonds can interact in turn in [3,3] sigmatropic rearrangements of the type previously discussed for homotropilidene. This molecule is therefore susceptible to extreme fluctuations. The ten carbon atoms in bullvalene can combine in $10!/3 = 1.2 \times 10^6$ different ways without changing the basic structure of bullvalene! Experiment bears out theory (see Problem V.1): The NMR spectrum of bullvalene at 100°C consists of a sharp singlet at δ 4.22 (line width 1.5 Hz). [For a detailed discussion, see G. Schröder, J. F. M. Oth, and R. Merényi, *Angew. Chem. Int. Ed. Engl.* **4**, 752 (1965).] Furthermore, base-catalyzed H–D exchange at a vinyl position in bullvalene has been shown to result in a statistical scrambling of deuterium over the entire molecule.

PROBLEM V.1

By means of successive [3,3] sigmatropic rearrangements, draw a sufficient number of representative structures to show that base-catalyzed H–D exchange at, e.g., the 10-position in bullvalene, can result in scrambling of the deuterium to every position in the molecule.

PROBLEM V.2

Bullvalene had now been prepared by several different routes. Many of the steps in these synthetic routes owe their success to the fact that orbital symmetry is conserved in concerted processes; these syntheses are of interest to us in this connection.

(a) The chemistry of the cyclooctatetraene dimer A has been extensively studied. It provides the best route presently available for large-scale (ca. 100 gm) preparation of bullvalene:

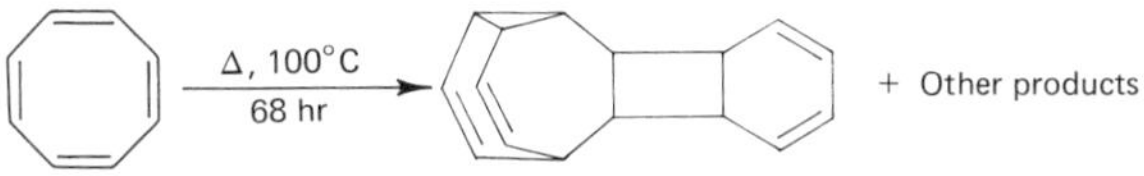

Δ, 100°C
68 hr
+ Other products
Dimer
(A)

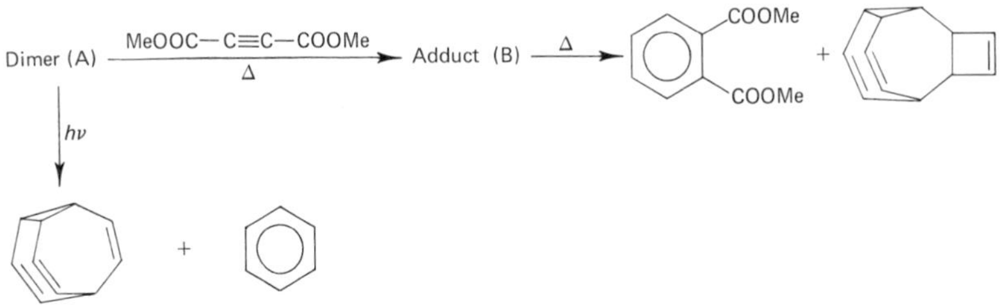

Dimer (A) — MeOOC—C≡C—COOMe, Δ → Adduct (B) — Δ → COOMe, COOMe +

hν

+

Bullvalene

Suggest a structure for adduct B and propose mechanisms for each of the two reactions of dimer A shown above.

(b) A second preparation of bullvalene is shown below:

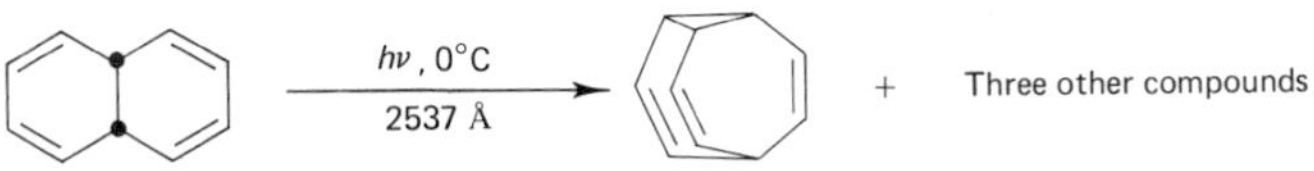

hν, 0°C
2537 Å
+ Three other compounds

In a later investigation, intermediate D was isolated in this photolysis.

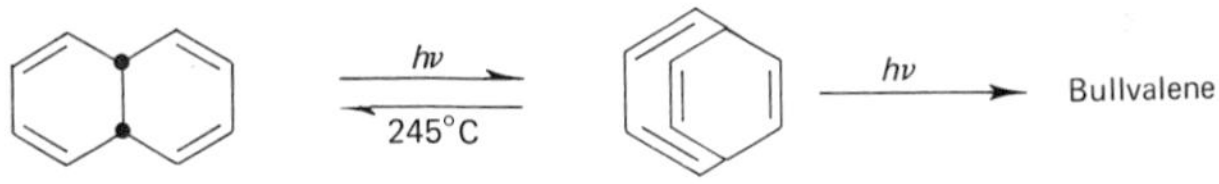

hν
245°C
(D)
hν
Bullvalene

Suggest a mechanism for the conversion of *cis*-9,10-dihydronaphthalene to bullvalene.

PROBLEM V.3

"Semibullvalene" (IV) possesses the homotropilidene structure (outlined in black in the figure below), and hence would be expected to display those properties associated with molecules possessing a fluctional structure.

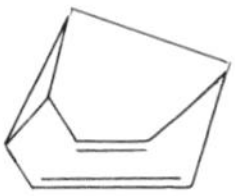

(IV)

(a) The NMR spectrum, which is unchanged in appearance when the temperature is varied from room temperature to $-110°C$, possesses the following characteristics:

Chemical shift	Area
4.92τ	2
5.83τ	4
7.03τ	2

Account for the appearance of the NMR spectrum of IV in terms of its anticipated fluctional structure.

(b) In a recent study, Professor H. E. Zimmerman at the University of Wisconsin has prepared semibullvalene via acetone-sensitized or by direct (unsensitized) photolysis of cyclooctatetraene.

$h\nu$, -40°C, acetone → IV + + (HC≡CH)

Potential intermediates in this reaction were tested, with the following results:

(1) 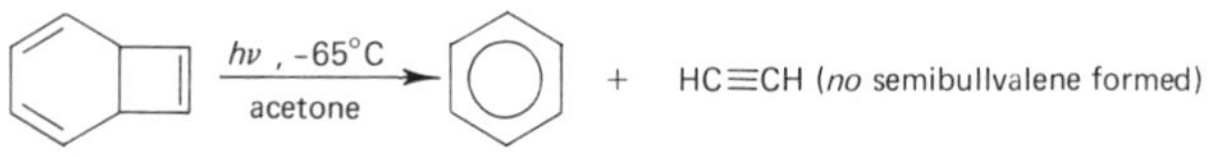 $h\nu$, -65°C, acetone → + HC≡CH (*no* semibullvalene formed)

(2) The following reaction is known:

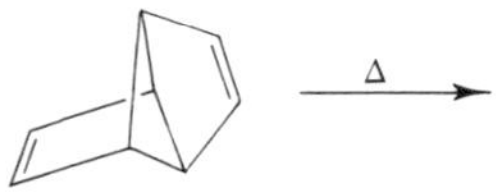

(V) $\xrightarrow{\Delta}$ (IV)

However, repeated attempts to establish the intermediacy of V in the photolytic conversion of cyclooctatetraene to semibullvalene failed. Suggest a mechanism for the rearrangement of cyclooctatetraene to semibullvalene under photolytic conditions in the presence of the acetone sensitizer. What is the function of the sensitizer?

(c) Semibullvalene can also be prepared by acetone-sensitized photolysis of bicyclo[2.2.2]octatriene ("barrelene," VI). It is also known that direct irradiation (no acetone present) of barrelene produces cyclooctatetraene. Direct irradiation of semibullvalene affords cyclooctatetraene but no barrelene. These reactions are summarized below:

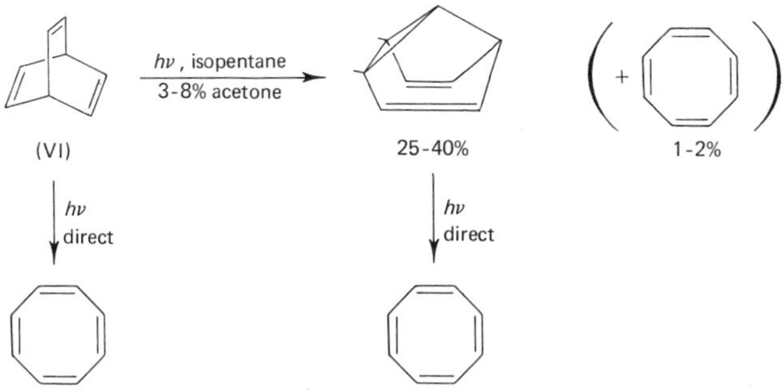

$h\nu$, isopentane
3-8% acetone

(VI) 25-40% 1-2%

$h\nu$ direct $h\nu$ direct

Account mechanistically for these results.

PROBLEM V.4

The photochemistry of bullvalene has been studied extensively by Professor M. Jones, Jr., at Princeton University. Two products of the direct irradiation of bullvalene, A and B, are of particular interest.

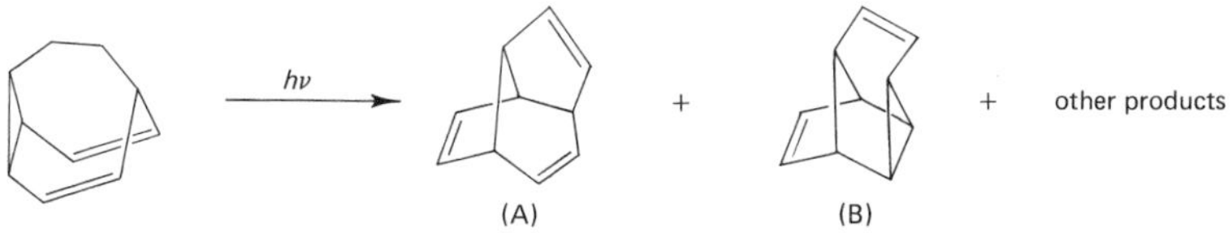

$h\nu$ + + other products

(A) (B)

Both A and B can be converted thermally to *cis*-9,10-dihydronaphthalene (C). The reaction with B is especially facile, occurring at 120°C in a few minutes.

Account mechanistically for the photochemical conversion of bullvalene to A and B, and for the thermal conversion of both A and B to *cis*-9,10-dihydronaphthalene (C).

Chapter VI

POTPOURRI

PROBLEM VI.1

Predict the products (show stereochemistry!) of each of the following reactions. Assume concerted reactions in all cases, which proceed to give the symmetry-allowed product.

(a) H_3C—[cyclic sulfone]—CH_3 $\xrightleftharpoons{\Delta}$ $SO_2 + C_6H_{10}$

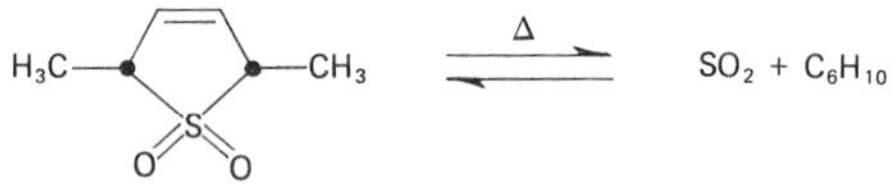

(b) [diketone] $\xrightarrow{h\nu}$ A nonconjugated polyolefin + a gas

(c) [benzonitrile, CN] + R—C≡C—R $\xrightarrow{h\nu}$ $C_9H_5NR_2$ [compound (A) contains *four* double bonds]

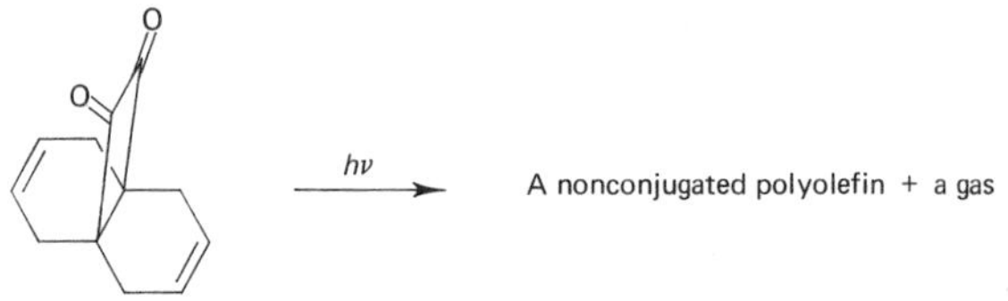

(A)

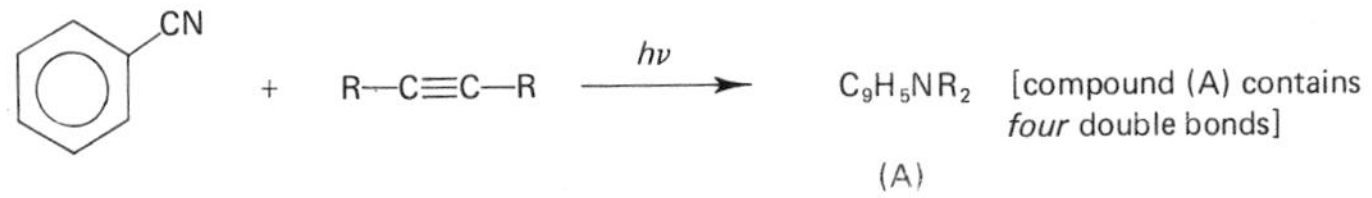

PROBLEM VI.2

Stable carbonium ion solutions in fluorosulfonic acid can be prepared. The results of a recent study in which stable cations in FSO_3H were subjected to photolysis at low temperature ($T < -60°C$) are shown below:

(a) [tropylium, +] BF_4^- (I) $\xrightarrow[h\nu,\ T < -60°C]{FSO_3H}$ [+] (II) $\xrightarrow[(+47°C)]{\Delta}$ [tropylium, +] (I)

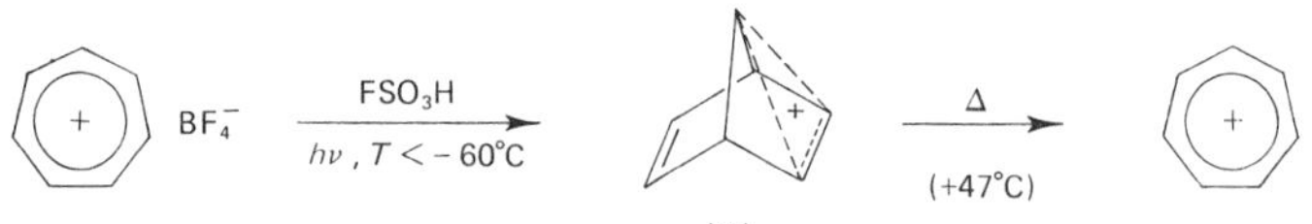

(b)

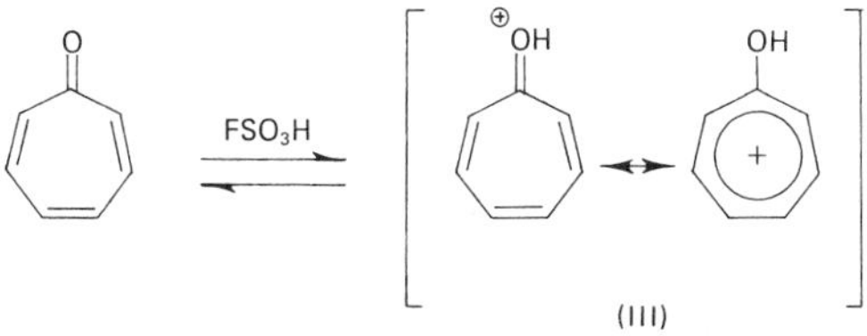

(III)

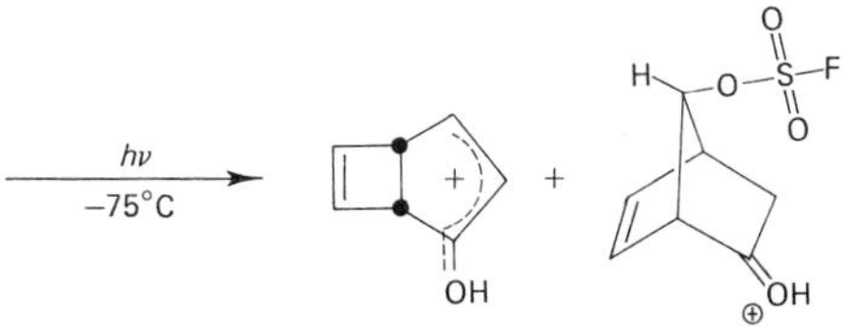

$\xrightarrow[-75°C]{h\nu}$

(IV) (V)

Note:

(1) The *thermal* reaction

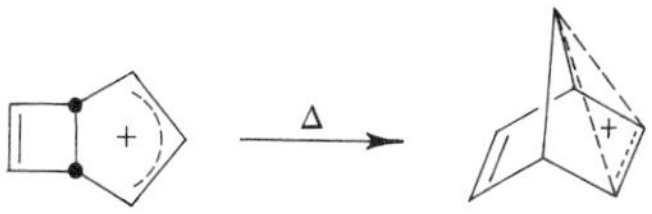

(VI) (II)

is known to be very rapid even at temperatures below − 60°C!

(2) Products IV and V are known to be stable in FSO_3H at low temperatures, (i.e., once formed they do not suffer further rearrangement).

Suggest mechanisms for reactions (a) and (b), above.

PROBLEM VI.3

Consider the following sequence of reactions:

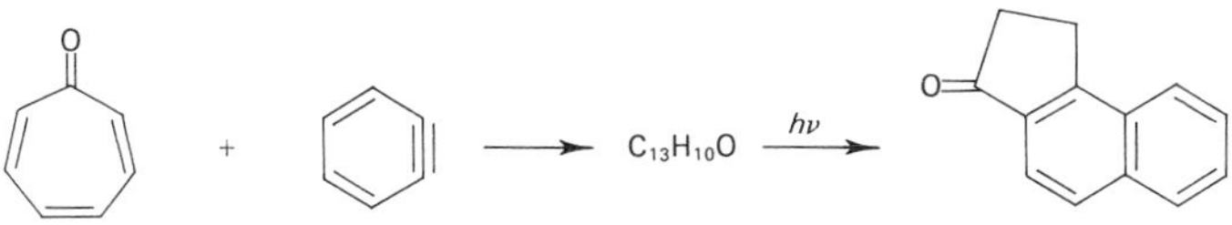

(generated *in situ*) $C_{13}H_{10}O$ (A) $\xrightarrow{h\nu}$ (B)

Suggest a structure for compound A, and suggest a mechanism for the photo-conversion A ⟶ B.

PROBLEM VI.4

Suggest a mechanism to explain the following observations:

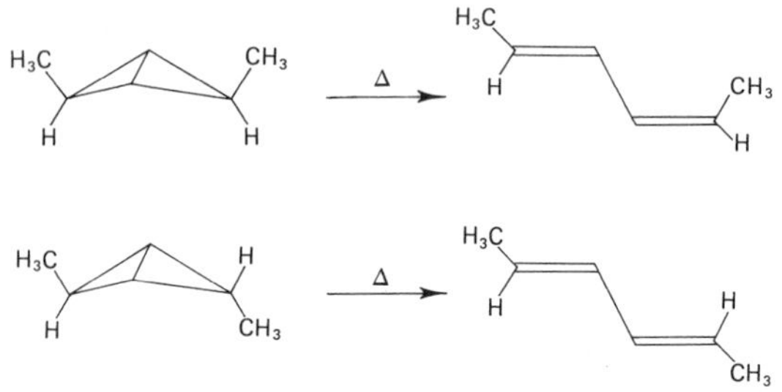

PROBLEM VI.5

Consider the following solvolytic processes:

(a) (I) $\xrightarrow[H_2O]{k_1}$ 2 CH_3CHO + NH_4Cl

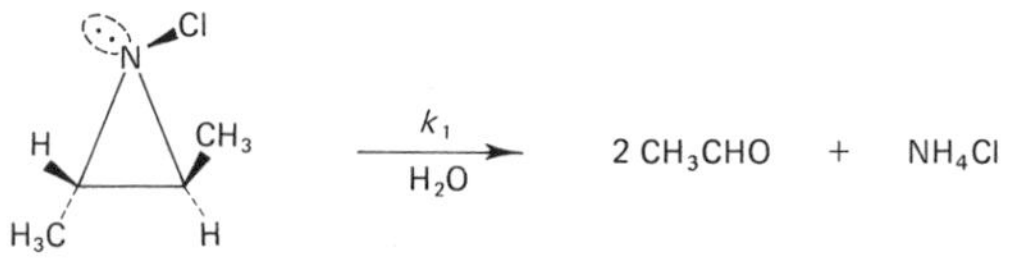

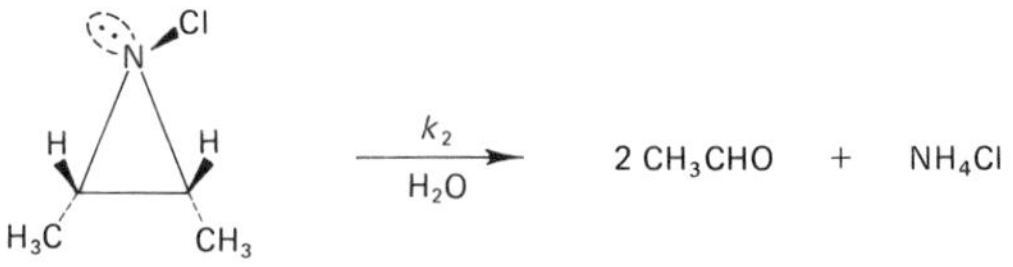

(b) (II) $\xrightarrow[H_2O]{k_2}$ 2 CH_3CHO + NH_4Cl

For these processes, the reactivity ratio k_2/k_1 = ca. 10^2. Suggest a mechanism for the solvolysis reaction of each *N*-chloroaziridine, and account for the reactivity difference between I and II.

PROBLEM VI.6

Suggest a mechanism for the following reaction:

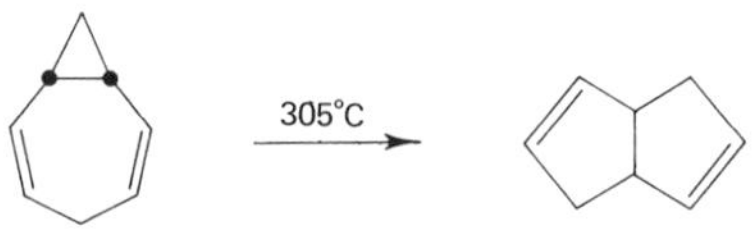

305°C

PROBLEM VI.7

Suggest a mechanism for the following reaction:

2 [cyclopentadiene] $\xrightarrow[\text{acetophenone (sensitizer)}]{h\nu}$ [cage dimer] + other products

PROBLEM VI.8

Suggest a mechanism for the following reaction:

[structure] $\xrightarrow{h\nu}$ [structure] ≡ [structure]

PROBLEM VI.9

Suggest a mechanism for the following reaction:

[1,2-bis(bromomethyl)cyclooctatetraene] + [bromocyclooctatetraene] $\xrightarrow[\text{Et}_2\text{O, 20°C}]{\text{KO-}t\text{-Bu}}$ [structure]

(I)

Of what theoretical interest is compound I?

PROBLEM VI.10

Suggest structures for compounds A and B, and suggest mechanisms for their formation. Compound A contains two double bonds, whereas compound B contains *no* double bonds.

[structure with HO, H] $\xrightarrow{h\nu}$ $C_9H_{10}O$ (A) $\xrightarrow[\text{acetone}]{h\nu}$ $C_9H_{10}O$ (B)

PROBLEM VI.11

2,3-Dimethyl-1,3-butadiene reacts with acetylenedicarboxylic acid under thermal conditions to produce a compound, $C_{16}H_{20}O_3$ (A), which contains two carbon–carbon double bonds. When A is photolyzed in acetone, a new compound, $C_{16}H_{20}O_3$ (B), is formed. Compound B contains *no* carbon–carbon double bonds. Suggest structures for A and B, and mechanisms for their formation.

CH$_3$ / CH$_3$ (2,3-dimethyl-1,3-butadiene) $\xrightarrow[\Delta,\ 76\%]{HO_2C\text{–}C{\equiv}C\text{–}CO_2H}$ $C_{16}H_{20}O_3$ (A) $\xrightarrow[\text{acetone}]{h\nu}$ $C_{16}H_{20}O_3$ (B)

PROBLEM VI.12

Suggest a mechanism for the following reaction:

(allyl cycloheptatrienyl ether) $\xrightarrow[24\text{ hr}]{200°C}$ (tricyclic ketone) + (tricyclic ketone)

PROBLEM VI.13

Bromocyclooctatetraene (I) has been found to rearrange almost quantitatively to *trans*-β-bromostyrene (A) at 80°C.

(I) $\xrightarrow{80°C}$ (A)

Huisgen has elegantly investigated the reaction and has uncovered the following information:

(a) The rearrangement, I⟶A, is 600 times faster if carried out in acetonitrile rather than cyclohexane as solvent. When the rearrangement is effected in the presence of LiI in acetone at 80°C, *trans*-β-iodostyrene is isolated in addition to A.

(b) When I is heated in the presence of 4-phenyl-1,2,4-triazoline-3,5-dione (II), the following four products are formed:

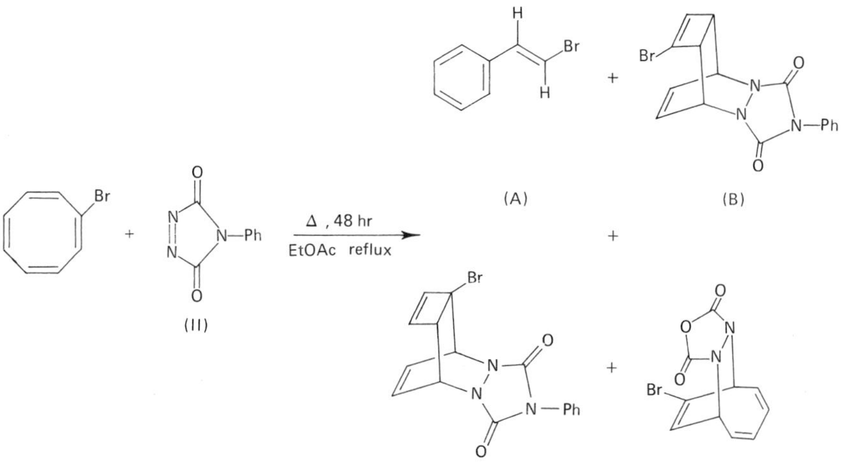

(II)

Δ, 48 hr
EtOAc reflux

(A) (B) (C) (D)

Propose mechanisms for the formation of compounds B, C, and D.
Propose a mechanism that accounts for the clean conversion of I to A in the absence of dienophile; your mechanism should also be consistent with the data in (a).
1,4-Dibromocyclooctatetraene (III) rearranges on heating to 180°C on an Apiezon L column to give a 92% yield of *p*-β-dibromostyrene (IV).

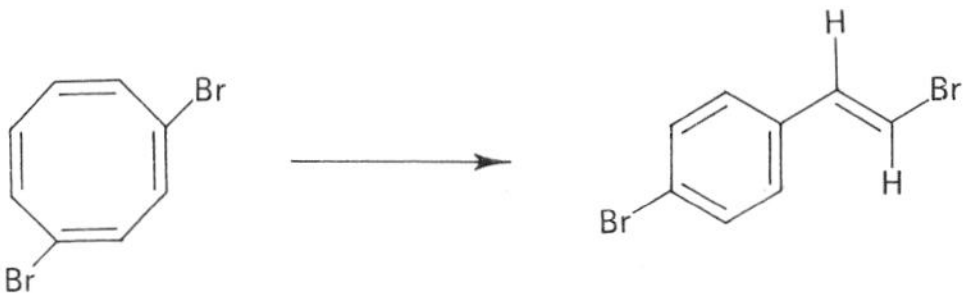

(III) (IV)

Suggest a mechanism for this reaction.

PROBLEM VI.14

Suggest a mechanism for the following reaction:

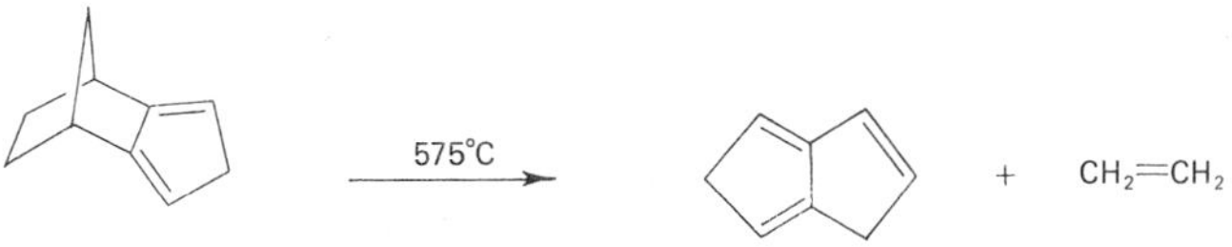

575°C

33%

+ $CH_2{=}CH_2$

PROBLEM VI.15

The following cheletropic process* has been investigated:

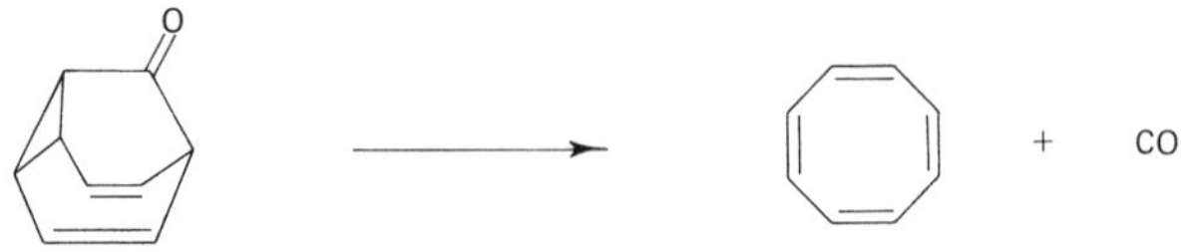

Assuming a *linear* transition state for this reaction, predict whether the concerted reaction is symmetry-allowed to occur thermally or photochemically. Suggest a mechanism for the symmetry-allowed reaction.

*T. Mukai and K. Kurabayashi, *J. Amer. Chem. Soc.* **92**, 4493 (1970).

Chapter VII

SYNTHESIS AND STRUCTURAL ANALYSIS

Stereospecific reactions are always of great interest to the synthetic chemist; through judicious selection, he is able to effect chemical conversions the stereochemical outcome of which is unequivocal. Structure proofs as well are often based on the ability or inability of a compound to smoothly undergo a reaction which is known to proceed stereospecifically, e.g., base-promoted *trans* E2 elimination of HX vs. *cis* elimination via xanthate pyrolysis (Chugaev reaction).

The remarkable stereospecificity of those kinetically controlled reactions which proceed with overall conservation of orbital symmetry suggests their application to structural analysis and stereospecific synthesis. The reader is afforded the opportunity to apply his understanding of orbital symmetry relationships in these ways in the following problems.

PROBLEM VII.1

Consider the flow diagrams (1) and (2):

(1) cyclooctatetraene $\underset{}{\overset{\Delta}{\rightleftharpoons}}$ $[C_8H_8]$ $\xrightarrow[MeO_2C-C\equiv C-CO_2Me]{\Delta}$ $C_{14}H_{14}O_4$ (A) $\xrightarrow[Pd/C]{D_2}$ $C_{14}H_{14}D_2O_4$ (B)

$\xrightarrow{200°C}$ $C_4H_4D_2$ (C) + dimethyl phthalate (CO_2Me, CO_2Me)

(C) $\xrightarrow{280°C}$ $C_4H_4D_2$ (D) $\xrightarrow[(2)\ H_2O]{(1)\ \text{maleic anhydride},\ \Delta}$ $C_8H_8D_2O_4$ (E) $\xrightarrow[\text{pyridine}]{Pb(OAc)_4}$ D〰(cyclohexadiene)〰D (F)

(F) $\xrightarrow[\text{stereospecific}]{\Delta}$ C_6H_5D (benzene–D) + HD

(2) [reaction scheme: 230°C → heat, nonstereospecific → D(H) + HD + D_2 + H_2]

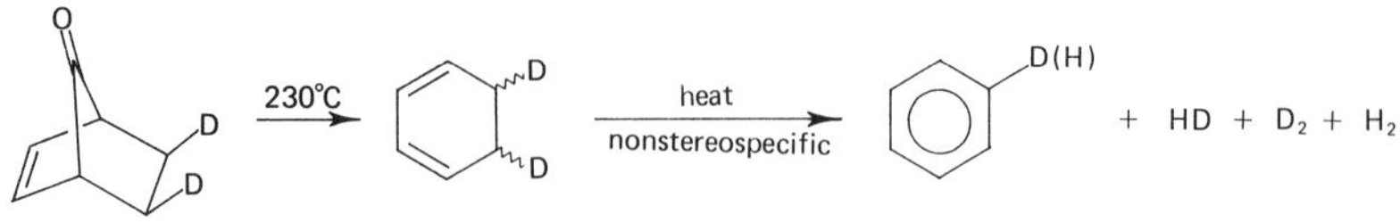

(a) Suggest structures (show stereochemistry) for the lettered products A through G.

(b) Explain why stereospecific loss of HD occurs upon pyrolysis of F, whereas the corresponding loss of molecular "hydrogen" from G occurs nonstereospecifically.

PROBLEM VII.2

(a) The following solvolysis reactions are known:

(1) 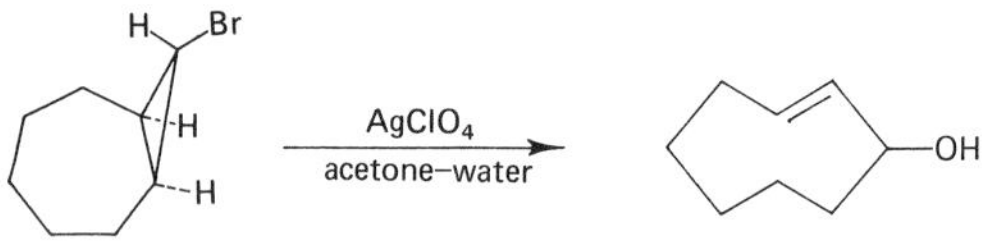 *exo*-Br —$AgClO_4$, acetone–water→ (*trans* double bond) –OH

(2) 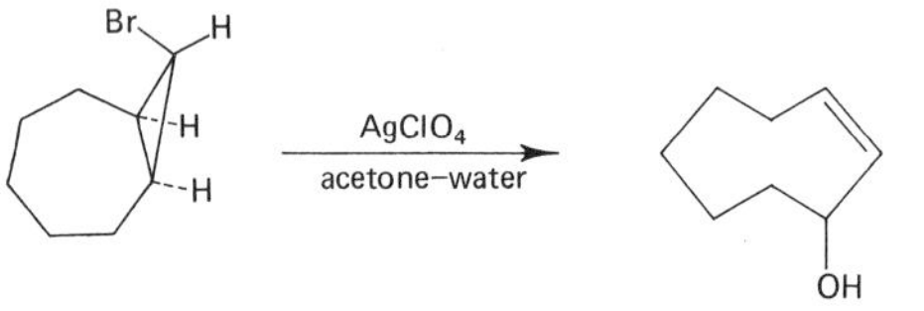 *endo*-Br —$AgClO_4$, acetone–water→ (*cis* double bond) OH

Suggest a mechanism for each of the two reactions indicated above. Which solvolysis reaction would be expected to occur more rapidly? Explain.

(b) The reaction of chlorobromocarbene with cyclohexene could conceivably lead to the formation of two products:

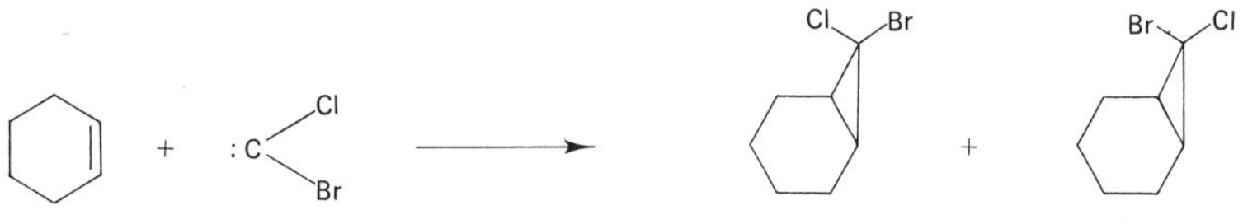

[cyclohexene + :C(Cl)(Br) → (A) + (B)]

Show how a structure proof of each of the two possible products, A and B, could be accomplished by examination of the products formed via simple solvolyses of A and B.

(c) Recently, Wiseman and co-workers* have prepared a series of compounds which "violate" Bredt's rule (i.e., compounds containing a bridgehead double bond):

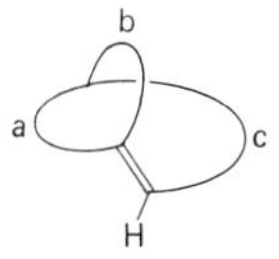

b
a c
H

Wiseman has argued that compounds of this type should be isolable when the larger of the two rings containing the endocyclic double bond (ac or bc) is at least eight-membered, and in which the endocyclic double bond is *trans*. Compounds of the following type meet these criteria:

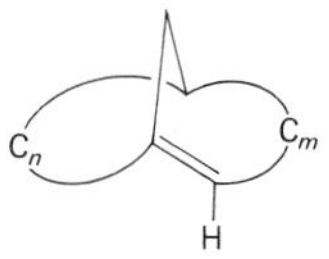

C_n C_m
H

With the above in mind and utilizing concepts inherent in the solution to parts (a) and (b) above, formulate a synthesis for compound C utilizing bicyclo[6.6.0]-Δ^{13}-tetradecene as starting material (assuming that this compound were readily available!).

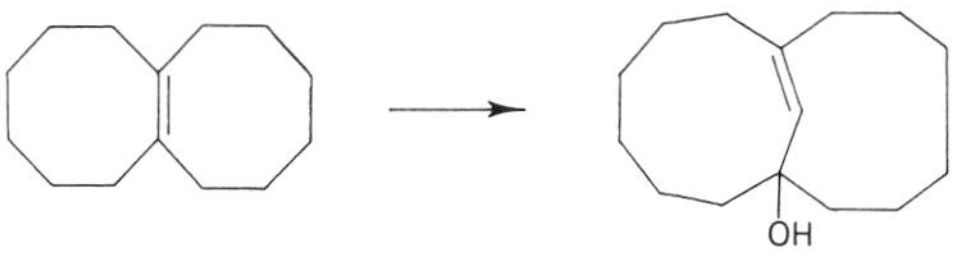

OH
(C)

PROBLEM VII.3

Cyclooctatetraene has proved to be a versatile starting material for the preparation of highly strained cage systems. One of these, a $C_{10}H_{10}$ isomer trivially named "basketene," has received considerable attention. The preparation of basketene is as follows†:

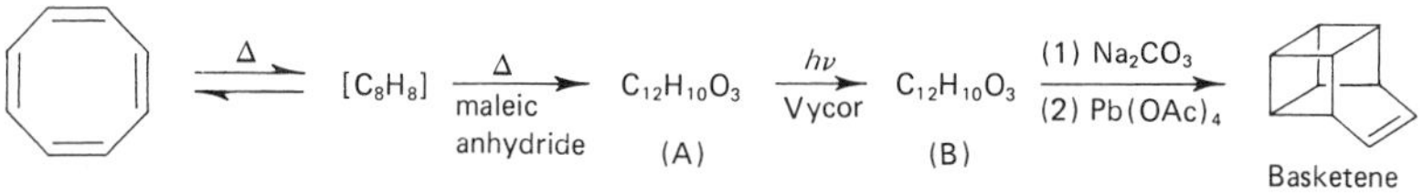

$\xrightleftharpoons{\Delta}$ $[C_8H_8]$ $\xrightarrow[\text{maleic anhydride}]{\Delta}$ $C_{12}H_{10}O_3$ (A) $\xrightarrow[\text{Vycor}]{h\nu}$ $C_{12}H_{10}O_3$ (B) $\xrightarrow[(2)\ Pb(OAc)_4]{(1)\ Na_2CO_3}$ Basketene

*See J. R. Wiseman, H.-F. Chan, and C. J. Ahola, *J. Amer. Chem. Soc.* **91**, 2812 (1969); J. R. Wiseman and J. A. Chong, *ibid.*, p. 7775; J. R. Wiseman and W. A. Pletcher, *ibid.* **92,** 956 (1970).

†S. Masamune, H. Cuts, and M. G. Hogben, *Tetrahedron Lett.* p. 1017 (1966).

Assign structures to compounds A and B.

The chemistry of basketene has been studied by Professor Eugene LeGoff at Michigan State University*:

Basketene $C_{10}H_{10}$ —(Δ, maleic anhydride)→ $C_{12}H_{10}O_3$ (A) + $C_{18}H_{14}O_6$ (C) —((1) MeOH, H^+; (2) $h\nu$)→ $C_{20}H_{26}O_8$ (E)

(A) —$h\nu$→ $C_{12}H_{10}O_3$ (G)

(A) —Δ→ $C_{12}H_{10}O_3$ (D)

(D) —Δ, maleic anhydride→ (C)

(D) —$h\nu$→ $C_{12}H_{10}O_3$ (F)

(G) —(1) H_2O; (2) $Pb(OAc)_4$→

Suggest structures for compounds A through G, and suggest detailed mechanisms for their formation.

PROBLEM VII.4

Professor Emanuel Vogel at the University of Cologne, Germany, has extensively studied the chemistry of a novel 10-pi-electron system, 1,6-methanocyclodecapentaene (I).

(I) (II)

The 1,6-methylene bridge permits a nearly planar (and hence, aromatic) configuration of the Hückel-aromatic conjugated 10-pi-electron system. This planar configuration cannot be attained by the parent *cis, cis, trans, cis, trans*-cyclodecapentaene (II) (note the proximity of the internal 1,6-carbon–hydrogen bonds).

The preparation of I was achieved via the *cis*-9,10-dihydronaphthalene derivative, III, which then spontaneously rearranged to I.

*E. LeGoff and S. Oka. *J. Amer. Chem. Soc.* **91**, 5665 (1969).

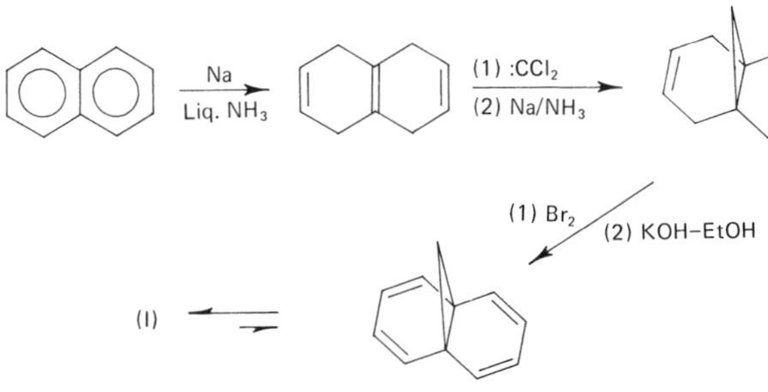

Na / Liq. NH_3; (1) $:CCl_2$ (2) Na/NH_3; (1) Br_2 (2) KOH–EtOH

(I) ⇌ (III)

(a) Account mechanistically (in orbital symmetry terms) for the ease of the thermal rearrangement III ⇌ I.

(b) Identify the lettered compounds A and B.

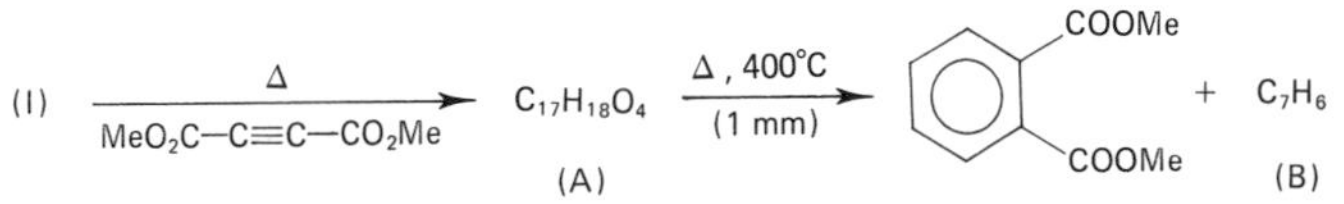

(I) $\xrightarrow[MeO_2C{-}C{\equiv}C{-}CO_2Me]{\Delta}$ $C_{17}H_{18}O_4$ (A) $\xrightarrow[(1\ mm)]{\Delta,\ 400°C}$ COOMe, COOMe + C_7H_6 (B)

(c) Attempts to prepare 11-oxo-1,6-methanocyclodecapentaene (IV) by oxidation of the corresponding alcohol V have not been successful. What products do you think were formed instead of IV? Account for the apparent instability of compound IV.

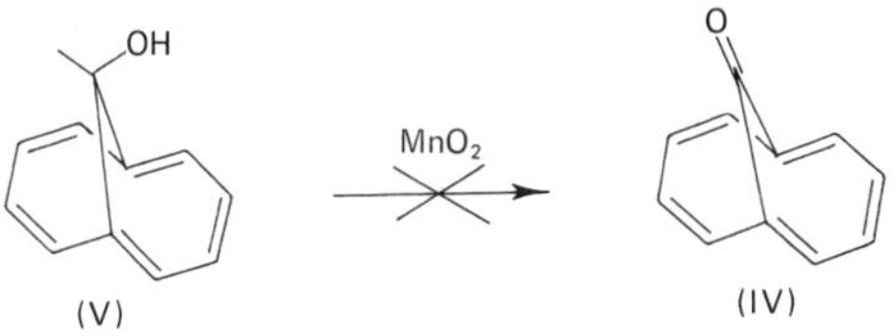

OH (V) —MnO_2 ✗→ O (IV)

(d) More recent studies by Vogel have led to the preparation of another 10-pi-electron system, VI; two methods for the preparation of compound VI are indicated below:

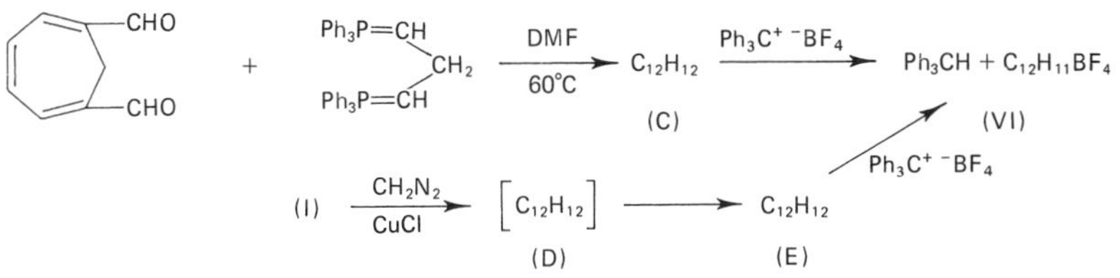

CHO, CHO + $Ph_3P{=}CH$, CH_2, $Ph_3P{=}CH$ $\xrightarrow[60°C]{DMF}$ $C_{12}H_{12}$ (C) $\xrightarrow{Ph_3C^+\ ^-BF_4}$ $Ph_3CH + C_{12}H_{11}BF_4$ (VI)

(I) $\xrightarrow[CuCl]{CH_2N_2}$ $[C_{12}H_{12}]$ (D) ⟶ $C_{12}H_{12}$ (E) $\xrightarrow{Ph_3C^+\ ^-BF_4}$ (VI)

Suggest structures for compounds VI and for the three $C_{12}H_{12}$ isomers, C, D, and E. Comment on the mechanism of the interconversion D $\rightleftharpoons$ E.

PROBLEM VII.5

(a) Consider the following reaction sequence:

$$\xrightarrow[\Delta]{SO_2} \xrightarrow{CrO_3} \xrightarrow{\Delta} SO_2 + C_7H_8O \text{ (A)} \xrightarrow[-196°C]{h\nu} C_6H_8 \text{ (B)}$$

Compound B is a highly unstable substance ($t_{1/2}$ = ca.20 min at − 196°C), and it displays a detailed EPR spectrum (observed simultaneously during irradiation) indicative of a molecule possessing a *triplet* (diradical) electronic state. Suggest structures for compounds A and B.

(b) Compound B, above, has been implicated in the degenerate thermal rearrangement of 1,2-dimethylenecyclobutane.

$$\xrightleftharpoons{275°C}$$ and

Suggest a mechanism for this reaction which postulates the intermediacy of B.

(c) A 1:1 molar mixture of C and D, after pyrolysis, gave a product containing *no* dideuterated material (as gleaned from mass spectral analysis).

(C) + (D) $\xrightarrow{275°C}$ Products (all $-d_0$ and $-d_4$)

What possible mechanism for the degenerate thermal rearrangement of 1,2-dimethylenecyclobutane is ruled out by this observation?

PROBLEM VII.6

Consider the following reaction sequences:

$$\xrightarrow[\text{room temp.}]{\text{Zn, DMF}} C_{10}H_{10} \text{ (A)} \xrightarrow[\Delta]{MeOOC-C\equiv C-COOMe} C_{16}H_{16}O_4 \text{ (B)}$$

(B) —[Cl, Cl, CN, CN quinone (DDQ); benzene, reflux]→ $C_{16}H_{14}O_4$ (C)

[bromocyclooctatetraene] —[KO-t-Bu, Et_2O]→ $[C_8H_6]$ (D) —[compound (A)]→ $C_{18}H_{16}$ (E) —[$-H_2$]→ $C_{18}H_{14}$ (F)

Suggest structures for compounds A through F.

PROBLEM VII.7

Although thermal Diels-Alder reactions are always stereospecific *cis* $[_{\pi}4_s + _{\pi}2_s]$ cycloadditions, a mixture of isomers can result from reaction of unsymmetrical dienes with unsymmetrical dienophiles. An example is the reaction of isoprene with methyl vinyl ketone.*

[isoprene] + [methyl vinyl ketone] —Δ→ (A) 29% + (B) 71%

Recently, Professor G. Büchi at Massachusetts Institute of Technology has developed a new method for the synthesis of cyclohexenes; in particular, compound A can be prepared in 75% yield, free from contamination by the isomer, B.

(a) Suggest a mechanism for the thermal rearrangement of the 3,4-dihydro-2*H*-pyranylethylene, C, to compound A.

(C) —Δ, 75%→ (A)

(b) Two possibilities exist for the geometry of the transition state involved in the reaction shown in part (a). What are these two possibilities? (*Hint:* See

*E. F. Lutz and G. M. Bailey, *J. Amer. Chem. Soc.* **86,** 3899 (1964).

Problem II.20.) Which of the two possibilities is actually operative in this case?

PROBLEM VII.8

Consider the following flow diagram:

MeOOC—C≡C—COOMe, Δ → $C_{14}H_{14}O_5$ (A) → (1) $NaBH_4$, MeOH; (2) Ac_2O, pyridine → $C_{16}H_{18}O_6$ (B)

(B) Δ → COOMe, COOMe + $C_6H_8O_2$ (C) + OAc

Assign structures to compounds A, B, and C, and comment briefly on the mechanism of formation of *trans*-1-acetoxybutadiene via pyrolysis of compound B.

Chapter VIII

PERICYCLIC REACTIONS: ADDENDA

Problems in this chapter were gathered from the 1971 chemical literature. No detailed answers to these questions have been supplied; instead, the original literature references to each problem have been given, and the reader is encouraged to compare his answers with those of the original authors.

PROBLEM VIII.1

The sodium salt of spiro[2.3]hexan-4-one tosylhydrazone (I) was pyrolyzed (gas phase, 350°C) and the products were collected in a liquid nitrogen-cooled trap. When the "pyrolysis products" were slowly warmed to room temperature, an exothermic reaction occurred from which compound A, C_6H_8, and a dimeric compound, B, $C_{12}H_{16}$, could be isolated. Dimer B was identified by gas-phase pyrolysis (350°C), which converted it into a new compound, C, whose structure is shown below.

When cyclopentadiene was added to the liquid nitrogen-cooled "pyrolysis products" *before* warming to room temperature, compound A along with a new hydrocarbon, D, could be isolated. No dimer B was found under these conditions.

Identify the compound(s) present in the "pyrolysis products," and suggest structures for compounds A and B. Suggest symmetry-allowed pathways for the formation of all compounds shown in the following scheme.

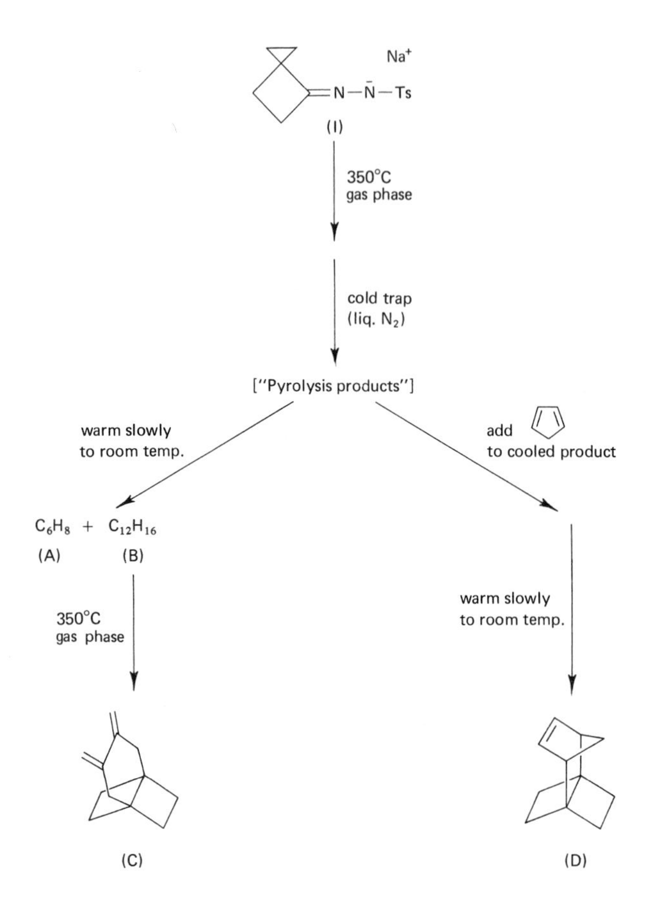

PROBLEM VIII.2

Suggest a detailed mechanism to account for the following transformation:

D D

Δ, gas phase

193°C, 33.5 hr

D D

no D D

(I)

(II)
(only product)

PROBLEM VIII.3

Consider the reaction shown below:

F, F, F, F, F, O + $H_2C{=}C{=}CH_2$ —(90°C, quantitative)→ F, F, F, F, F, OH, $CH_2{-}C{\equiv}CH$

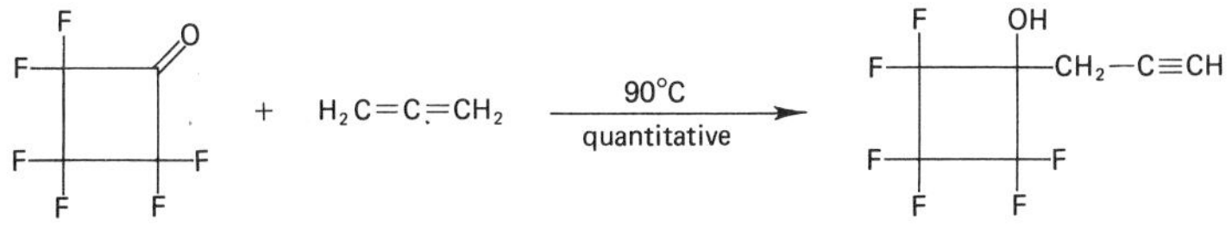

The following kinetic isotope effects, k_H/k_D, were observed:

Reactant	k_H/k_D
$D_2C{=}C{=}CH_2$	3.53 ± 0.03
$DHC{=}C{=}CHD$	2.35 ± 0.05

Suggest a mechanism for this reaction which is consistent with the observed k_H/k_D values.

PROBLEM VIII.4

(a) On the basis of orbital symmetry considerations, predict the stereochemistry of the *phenyl* group in the product of the reaction shown below:

Ph, H, H, Me, CN, CN —($h\nu$)→ Ph, CN, CN, H, Me, CH_2

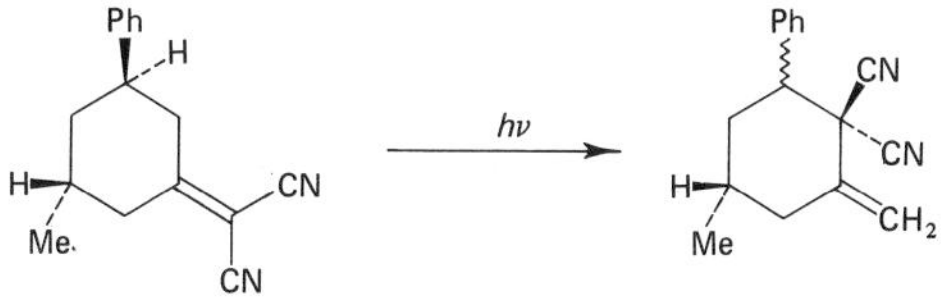

(b) On the basis of orbital symmetry considerations, predict the stereochemistry of the *methyl* group in the product of the reaction shown below:

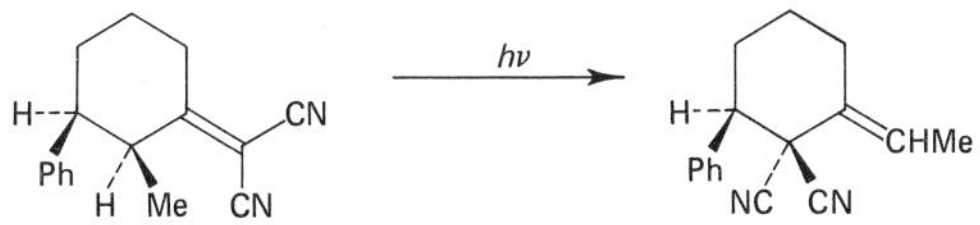

H, Ph, H, Me, CN, CN —($h\nu$)→ H, Ph, NC, CN, CHMe (Me *cis* or *trans*?)

PROBLEM VIII.5

Suggest a symmetry-allowed pathway to account for the following transformation:

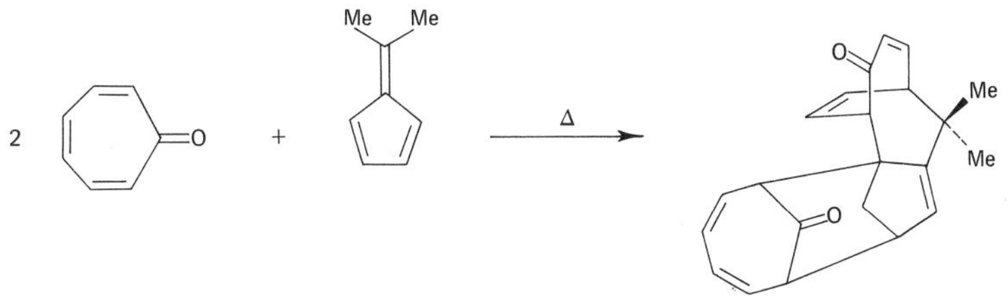

2 (O) + (Me, Me) —(Δ)→ (O, Me, Me, O)

PROBLEM VIII.6

The following reaction has been observed:

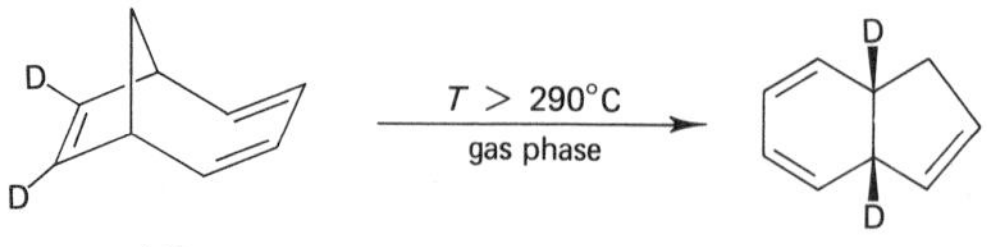

$T > 290°C$, gas phase

(A) (B)

A possible intermediate in this reaction, C, has been tested successfully:

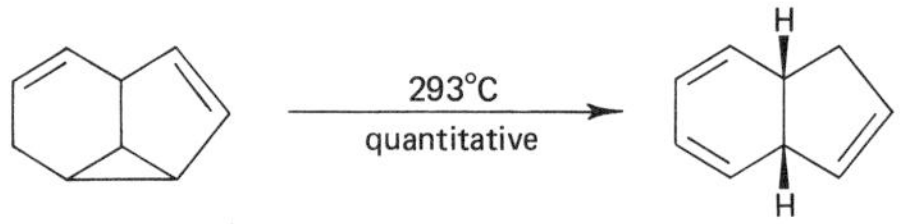

293°C, quantitative

(C)

Account mechanistically for the thermal conversion A ⟶ B, invoking the intermediacy of C. Individual (concerted) steps in your mechanism should be in accord with predictions based on orbital symmetry considerations.

PROBLEM VIII.7

The thermal transformation of 9,10-dideuteriosnoutene (A) to 7,8-dideuteriosnoutene (B) has been studied.

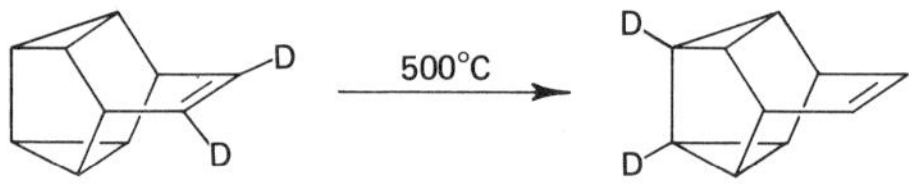

500°C

(A) (B)

This degenerate rearrangement of snoutene is extremely specific; *only* the indicated vinyl and cyclopropyl protons are scrambled at 500°C. However, at 530°C, A is smoothly transformed to *cis*-9,10-dihydronaphthalene-d_2, C, in which the two deuterium atoms have become statistically scrambled throughout the molecule.

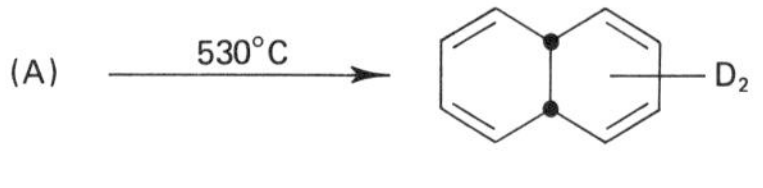

(A) 530°C → D_2

(C)

Suggest one or a series of concerted steps for each of the transformations A ⟶ B and A ⟶ C which are in accord with predictions based on orbital symmetry considerations.

PROBLEM VIII.8

Suggest a mechanism to account for the following transformation:

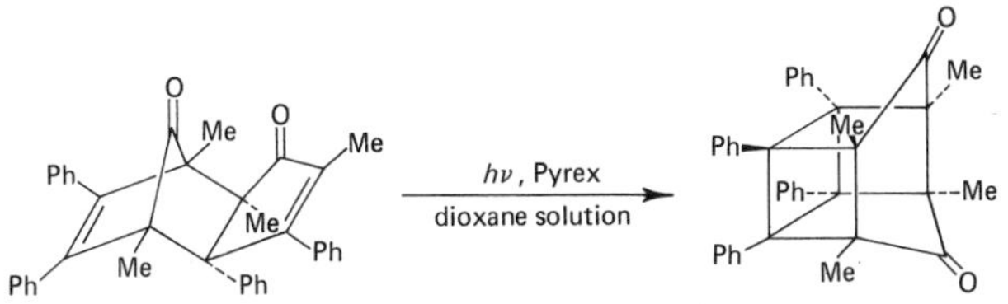

PROBLEM VIII.9

The photoisomerization of optically active 1,1-dicyano-2-methyl-4-phenylpent-1-ene, A, to 3,3-dicyano-2-methyl-4-phenylpent-1-ene, B, proceeds with ~85% retention and 15% inversion of configuration at the migrating center:

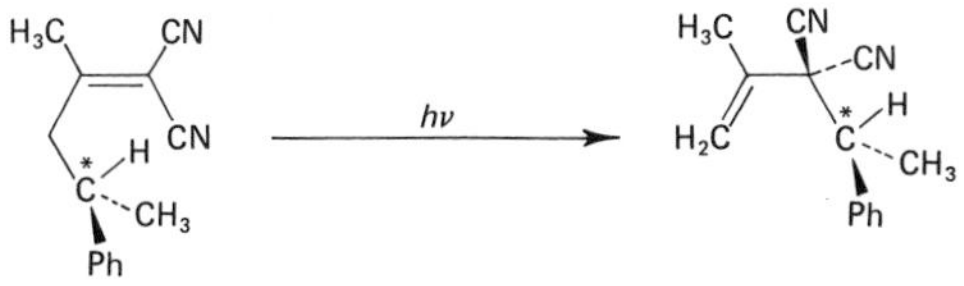

However, *thermal* reversion of this reaction (i.e., B $\xrightarrow{\Delta}$ A) occurs with >90% *retention* of configuration at the migrating center. Suggest concerted, symmetry-allowed pathways for each reaction, and comment briefly in each case as to whether you believe your mechanism is energetically and sterically feasible.

PROBLEM VIII.10

In each case, offer a detailed explanation to account for the following observations:

(a) Direct photolysis of benzene in D_3PO_4–D_2O affords *only exo*-2-hydroxy-*endo*-6-deuteriobicyclo[3.1.0]hex-3-ene, A; *all* of the deuterium is incorporated into the 6-*endo* position of A.

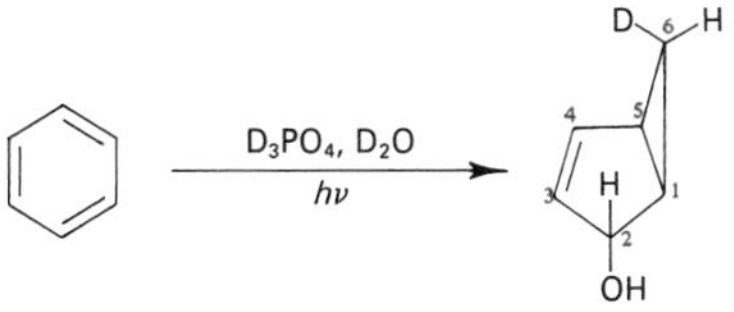

(b) Storage of solutions of the monodeuterated cation, B, at $-90°$ results in proton–deuterium equilibration throughout the molecule according to the following pattern: *proton* ratio $H_{1,5}:H_{2,4}:H_{3}:H_{6\text{-}endo}:H_{6\text{-}exo} = 1.6:1.6:0.8:1:1$.

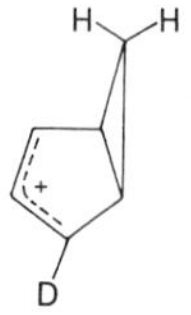

(B) 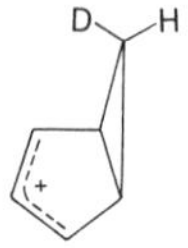 (C)

Note: Experiments with the 6-*endo*-deuterated cation, C, reveal that the process responsible for H–D equilibration in B does *not* concomitantly destroy the stereochemical integrity of the 6-*endo* and 6-*exo* positions. (*Explain*!)

PROBLEM VIII.11

In any dynamic and rapidly developing field, new discoveries are constantly unfolding which provide unforeseen paths into hitherto unexplored areas. As this book goes to press, such a development promises to dramatically extend the scope of applications of pericyclic reaction chemistry: transition metals such as Ag(I), Rh(I), Pt(IV), Pd(II), Ni(III), and Ir(I) as well as other transition metal elements have been found to catalyze pericyclic reactions in unusual ways. Examples of this effect abound, but explanations which have been forwarded to account for the observed effects of transition metal catalysis are conflicting, and there appears to be little basis at present to generally favor one explanation over another. It is therefore fitting and proper that our book should end on a question mark. In the examples which follow, we present some representative results from the current literature; the challenge to the reader is to piece together the individual puzzle parts and to construct a comprehensive theory.

(a) Classify each of the following reactions as $[m + n]$ pericyclic processes. State in each case whether the indicated reaction, if proceeding in a concerted fashion, is in accord with prediction based on orbital symmetry considerations. (*Disregard* the presence of the transition metal catalyst for the purpose of assessing the symmetry-allowedness of these examples.)

(1) 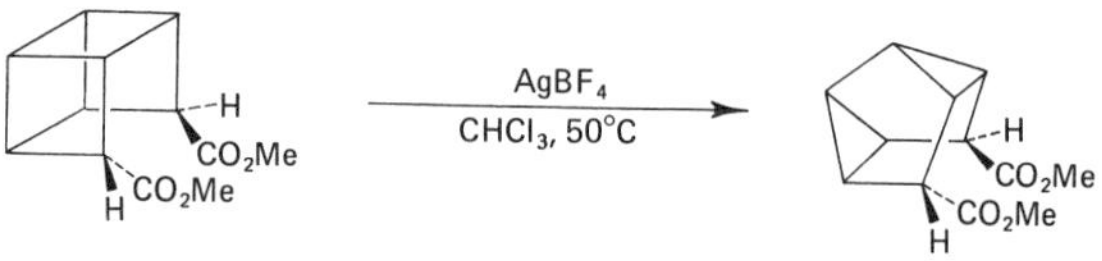 $\xrightarrow[\text{CHCl}_3,\ 50°\text{C}]{\text{AgBF}_4}$

(2) 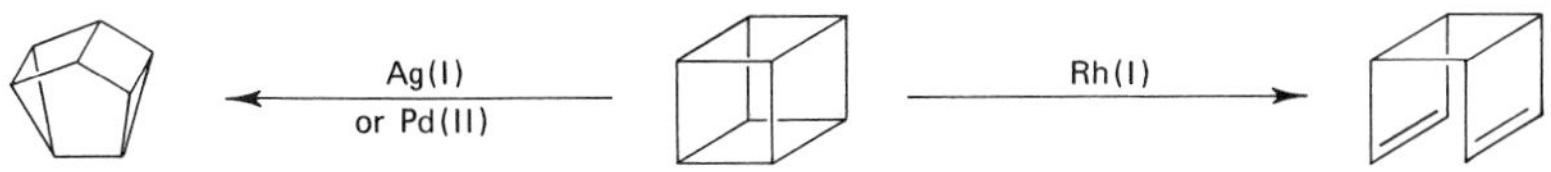 Ag(I) or Pd(II); Rh(I)

(3) 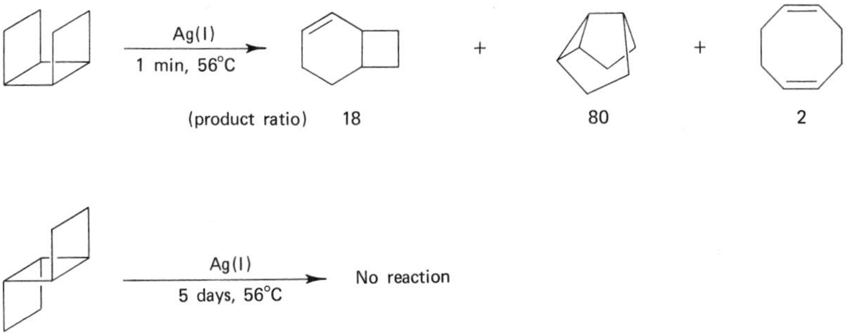 Ag(I), 1 min, 56°C

(product ratio) 18 + 80 + 2

Ag(I), 5 days, 56°C — No reaction

(4) 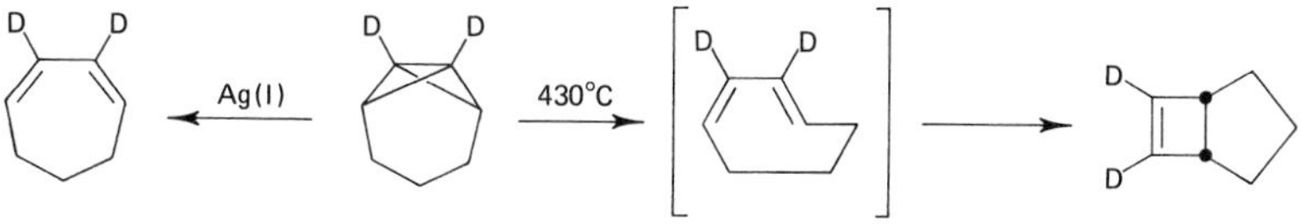 Ag(I); 430°C

D D D D D D D D

(5) 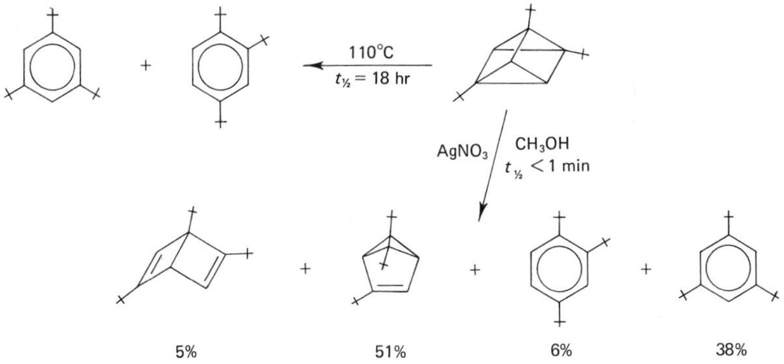 110°C, $t_{1/2} = 18$ hr

$AgNO_3$ / CH_3OH, $t_{1/2} < 1$ min

5% + 51% + 6% + 38%

(6) 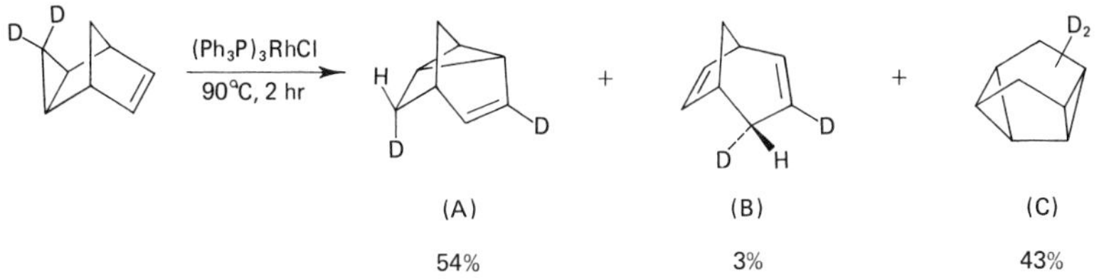 $(Ph_3P)_3RhCl$, 90°C, 2 hr

D D H D D D D H D_2

(A) 54% + (B) 3% + (C) 43%

Which positions would you expect the two deuterium atoms in C to occupy? *Explain.*

(b) Several investigators have raised the issue that the outcome of the competition between symmetry-allowed and -forbidden pathways is often prejudiced by geometrical factors inherent in the systems studied. This is particularly true of many of the highly strained, rigid ring systems which have commonly been chosen for study. Thus, the consonance (or dissonance) of a given pericyclic reaction with the principles of conservation of orbital symmetry may be fortuitous, the outcome being predetermined by geometric and/or steric factors which are extraneous to these rules.

The bicyclo[1.1.0]butane system is especially interesting in this regard, as no overwhelming steric factor is present which might unduly favor the symmetry-allowed over the symmetry-disallowed process, or vice versa (see Problem VI.4).

(1) The results given have been obtained for the uncatalyzed and Ag(I)-catalyzed thermal rearrangement shown below:

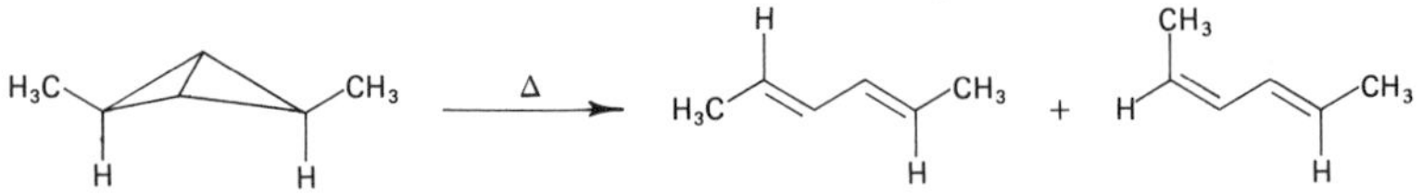

H_3C, H, H, CH_3 → (Δ) H_3C, H, CH_3, H + H, CH_3, CH_3, H

(A) (B) (C)

Conditions	%Yield, B	%Yield, C
200°C, uncatalyzed	3.9	93.2
Ag(I)—catalyzed, 26°C	77.0	23.0

Comment briefly on the effect of Ag(I) catalysis on the course of this rearrangement. Are the results here consistent with those observed for the Ag(I) catalyzed examples shown in part (a) of this question?

(2) In the preceding example, we have seen that the course of the rearrangement was essentially reversed upon introduction of the Ag(I) catalyst. However, the situation is certainly more complicated than such a statement might imply. Consider the following example:

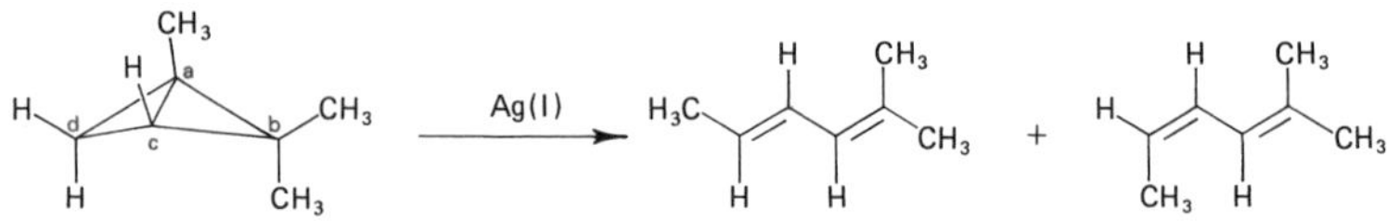

CH_3, H, a, H, d, c, b, CH_3, H, CH_3 → (Ag(I)) H_3C, H, CH_3, CH_3, H, H + H, H, CH_3, CH_3, CH_3, H

(D) 4.5 : 1 (product ratio)

What combination of bonds (a—b, b—c, a—c, c—d, and/or a—d) is broken in the Ag(I)-catalyzed rearrangement of D? Is a concerted or a stepwise mechanism implied for this rearrangement? Compare your answer with the discussion given in *J. Amer. Chem. Soc.* **93**, 2336 (1971).

(3) Consider the following rearrangements:

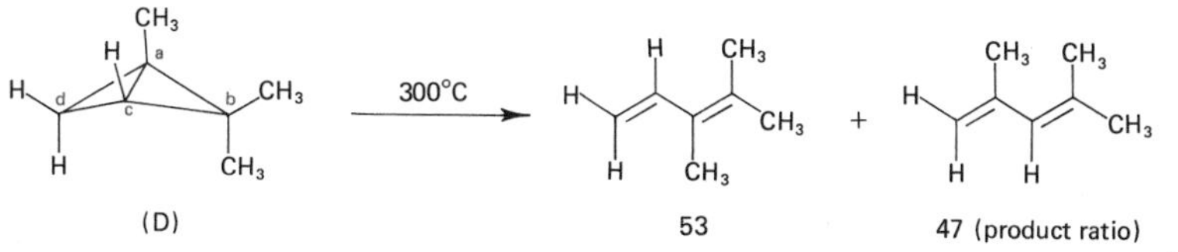

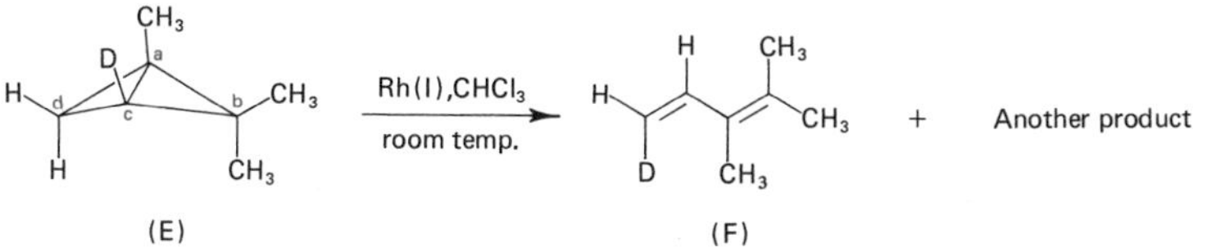

(D) —300°C→ 53 + 47 (product ratio)

(E) —Rh(I), $CHCl_3$, room temp.→ (F) + Another product

What combination of bonds (a—b, b—c, a—c, c—d, and/or a—d) is broken (i) in the thermal rearrangement of D, and (ii) in the Rh(I)-catalyzed rearrangement of E (which affords F as the *only diene* formed)? Can you suggest a mechanism for the conversion E ⟶ F? Is a concerted or a stepwise mechanism implied for this rearrangement? Can you suggest any reason why the Rh(I)-catalyzed rearrangement of E produces only one diene (F) in a highly stereospecific process, in contrast to the Ag(I)-catalyzed rearrangement of D (preceding part of this problem)?

(4) Additional alkyl substitution has been found to have a dramatic effect upon the course of transition metal-catalyzed thermal rearrangements of bicyclo[1.1.0]butane. Consider the following examples:

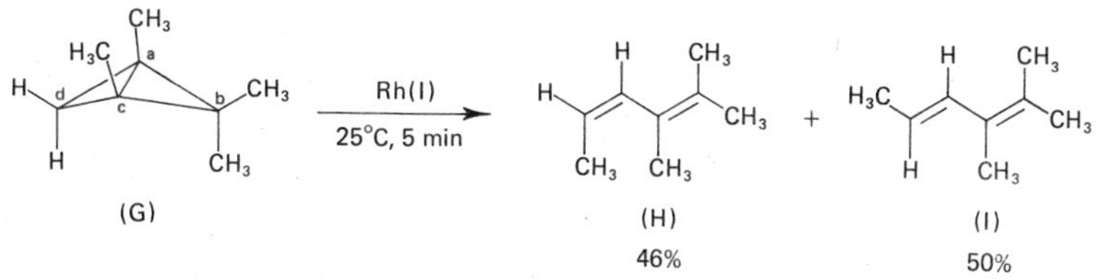

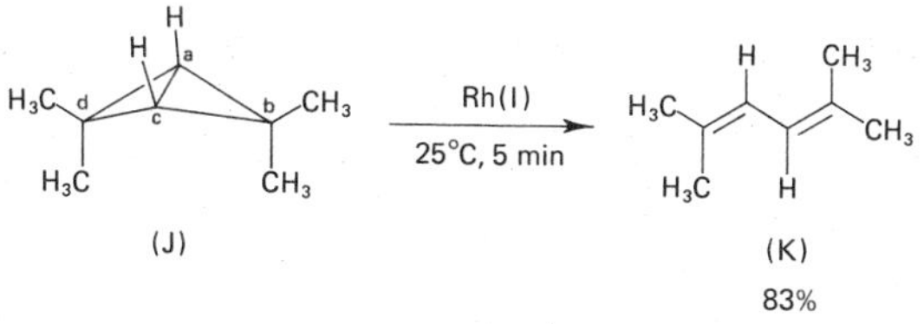

(G) —Rh(I), 25°C, 5 min→ (H) 46% + (I) 50%

(J) —Rh(I), 25°C, 5 min→ (K) 83%

What combination of bonds (a—b, b—c, a—c, c—d, and/or a—d) is broken in (i) the Rh(I)-catalyzed rearrangement of G, and (ii) the corresponding rearrangement of J? Compare the Rh(I)-catalyzed rearrangements of G and J with that of E in the preceding part of this problem. Is a concerted or a stepwise mechanism implied (i) for the Rh(I)-catalyzed thermal rearrangement of G, and (ii) for the corresponding rearrangement of J? Compare your answer with the discussion given in *J. Amer. Chem. Soc.* **93**, 1812 (1971).

(c) The following examples from the current literature utilize transition metal-catalyzed pericyclic reactions in synthetic schemes.

(1) Show preparation of A from readily available materials.

$\longrightarrow \longrightarrow$ (A) [structure with CO_2Me, CO_2Me]

(A) $\xrightarrow{h\nu}$ $\xrightarrow{Ag(I)}$ $\xrightarrow[2.\ Pb(OAc)_4]{1.\ OH^-}$ $C_{10}H_{10}$ (B) $\xleftarrow{Ag(I)}$ [structure]

Compound B contains one carbon–carbon double bond. Suggest a structure for compound B which is consistent with the information given above.

(2) Show preparation of C from readily available materials.

$\longrightarrow \longrightarrow$ (C) [structure with N, N, N–Ph, O, O]

(C) $\xrightarrow[acetone]{h\nu}$ $C_{16}H_{13}N_3O_2$ (D) $\xrightarrow{Ag(I)}$ $C_{16}H_{13}N_3O_2$ (E) $\xrightarrow[HOCH_2CH_2OH]{KOH,\ H_2O}$ $[C_8H_8N_2]$ (F) Not isolated $\xrightarrow{-N_2}$ (G)

Suggest structures for compounds D and E. Suggest a mechanism for the conversion of intermediate F to semibullvalene, G.

(3)

[cyclobutadiene–$Fe(CO)_3$] $\xrightarrow{Ce(III)}$ [[cyclobutadiene]] $\xrightarrow[\Delta]{O=C_6H_4=O}$ $C_{10}H_8O_2$ (H) $\xrightarrow{h\nu}$ $C_{10}H_8O_2$ (I) $\xrightarrow{Rh(I)}$ (J) [structure with O, O]

Compound H contains two carbon–carbon double bonds, but compound I contains none. Suggest structures for compounds H and I which are consistent with the information given above.

Appendix A

MOLECULAR ORBITAL THEORY

We begin our discussion with a familiar concept, *hybridization*. Thus, butadiene, $H_2C{=}CH{-}CH{=}CH_2$, is composed of four sp^2 (trigonally)-hybridized carbon atoms which together with six hydrogen atoms constitute the familiar sigma-bonded framework depicted in Fig. A.1.

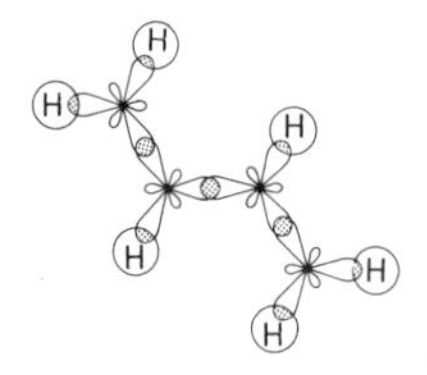

Fig. A.1. *Sigma-bonded framework of butadiene (shown in the s-trans conformation).*

In addition to the orbitals described above, each carbon atom possesses a $2p$ atomic orbital which lies orthogonal (i.e., at right angles to) the plane described by the sigma-bonded framework (Fig. A.2); each $2p_z$ atomic orbital of butadiene contains one electron. Although the bonds in the sigma-framework contribute substantially to the total energy of the molecule, those properties relating to the chemical reactivity of butadiene can best be described in terms of the properties

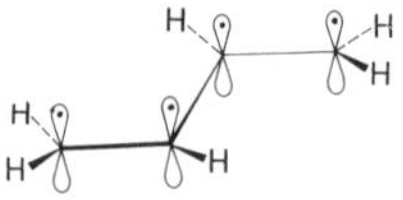

Fig. A.2. *$2p_z$ Atomic orbitals of butadiene.*

of the pi-electron system (i.e., electrons in the $2p_z$ atomic orbitals). For example, 1,4-conjugate addition (as in the addition of a dienophile to butadiene in the familiar Diels-Alder reaction), as well as 1,2-addition to butadiene, is commonly observed. Thus, as a first approximation, it is common practice to disregard the very stable sigma-bonded framework of butadiene as invariant (i.e., "localized") and to account for the chemical properties along with other collective or "many-electron" properties of butadiene in terms of the properties of the pi-electron system.*

When dealing with unsaturated and/or conjugated pi-systems, we focus our attention on the pi-electron systems. In the molecular orbital (MO) method, we combine the n $2p_z$ atomic orbitals (AO's) to produce n new MO's; in the case of butadiene, four AO's combine to form four new MO's. In the simplest MO method, that originated by Hückel in 1931,† this is accomplished by "linear combination of atomic orbitals" (the LCAO-MO method). Thus, for a given electron in the MO, the wave function, ψ, is simply the sum of the wave functions, ϕ, of the contributing AO's, each multiplied by an appropriate coefficient, c. For the jth molecular orbital, this becomes

$$\psi_j = c_{j1}\phi_1 + c_{j2}\phi_2 + c_{j3}\phi_3 + \cdots + c_{ji}\phi_i$$

$$= \sum_{j=1}^{n} c_{ji}\phi_i \qquad \text{(summed over } n \text{ AO's)} \tag{A.1}$$

where c_{ji} is the coefficient of the ith atomic orbital (ϕ_i) in the jth molecular orbital. The wave function, ψ, has the physical significance that when multiplied by its complex conjugate, ψ^*, and normalized, it represents the probability of finding an electron in the incremental volume element, $d\tau$ (i.e., $\psi\psi^*$ is a point electron density). Normalization simply means that the integral $\int \psi\psi^* d\tau = 1$; the probability of finding the electron in all space is 100%.

Solution of the wave function is properly obtained via the Schrödinger wave equation, through which discrete energies can be assigned to each of the wave functions, ψ. However, explicit formulation of the Schrödinger wave equation leads to intractibly complex expressions for any system more complicated than the hydrogen atom! Hückel's great contribution was to recognize that approximate solutions to the Schrödinger wave equation can be obtained utilizing the LCAO approach†; an n pi-electron system becomes n one-electron problems, the whole being the sum of its interacting parts.

A number of excellent treatments are available which present the Hückel MO method as well as higher-order approximations to solving the Schrödinger wave

* M. J. S. Dewar, "The Molecule Orbital Theory of Organic Chemistry," Chapter 4. McGraw-Hill, New York, 1969.

† E. Hückel, *Z. Physik* **70**, 204 (1931).

equation in polyatomic systems; the interested reader is referred to the Bibliography at the end of Appendix A. To satisfy our immediate needs, it is necessary only that we have in hand the solutions themselves, i.e., the magnitudes and algebraic signs of the AO coefficients, c_{ji}, and the energies, ϵ_j, of the individual MO's, ψ_j. With these results in hand, we can explore their significance in determining the outcome of concerted reactions.

Let us examine a typical set of MO's, ψ_j, and their corresponding energies, ϵ_j, those associated with our example, butadiene (Fig. A.3). The energies, ϵ_j, are

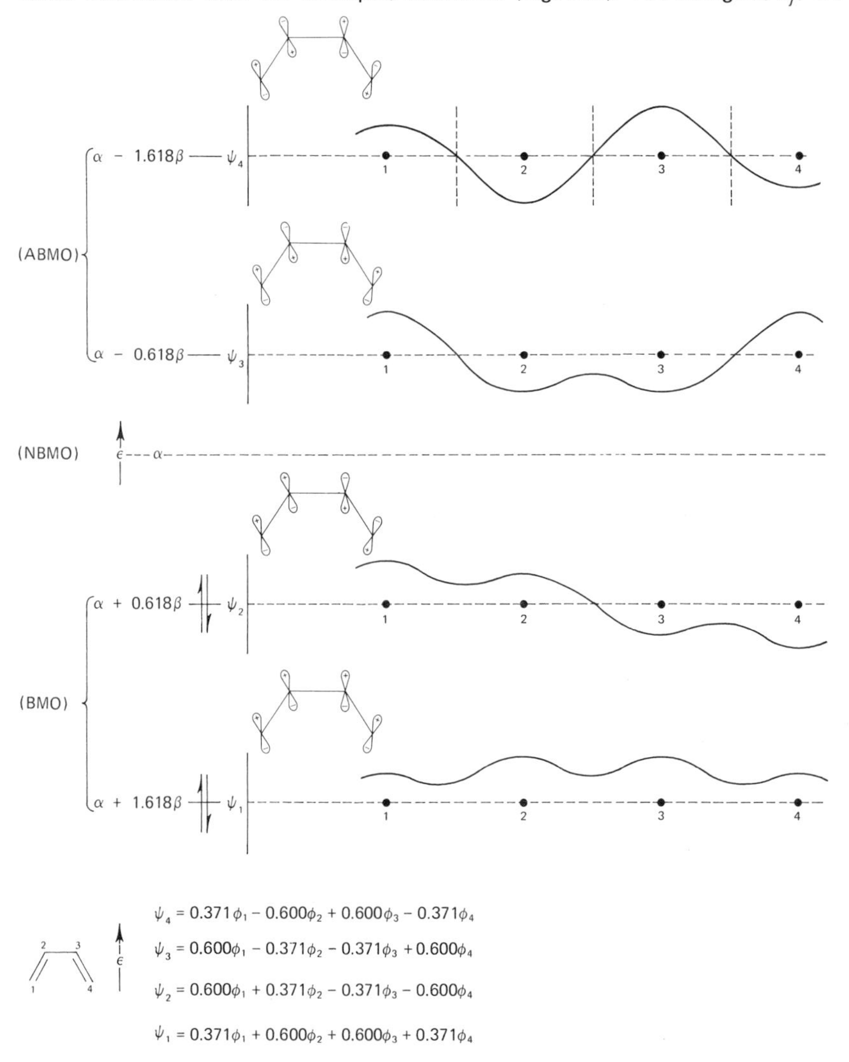

Fig. A.3. *Molecular orbitals of butadiene.*

commonly expressed in the form $\epsilon_j = \alpha_r + m_j\beta_{rs}$, where the parameters α and β represent the *coulomb integral* and the *resonance integral*, respectively. The coulomb integral of atom r, α_r, represents the ground-state ionization potential of an electron occupying the AO of the isolated atom, r. It may be generally thought of as a measure of the electronegativity of the isolated atom, r, and it is a negative number as defined. Thus, sp^2-hybridized nitrogen, a more electronegative atom than sp^2-hybridized carbon, will have the more negative (greater absolute) value of α_r.

The resonance integral, β_{rs}, represents the energy of interaction of the AO's associated with contiguous atoms r and s. This quantity as defined is likewise a negative number, and it is a measure of the *bond strength* (and, hence, of *bond length* and *bond order*) of the bond linking atoms r and s. For contiguous atoms, $\beta_{rs} > 0$, but for nonbonded atoms, $\beta_{rs} = 0$.

Finally, the value of m_j is significant. For $m_j > 0$, ψ_j is said to be a *bonding molecular orbital* (BMO); for $m_j < 0$, ψ_j becomes an *antibonding molecular orbital* (ABMO). For the special case $m_j = 0$, then $\epsilon_j = \alpha$, and ψ_j is said to be a nonbonding molecular orbital (NBMO). The significance of these three classifications of MO's can be appreciated from consideration of Fig. A.3.

Inspection of Fig. A.3 reveals that butadiene possesses two BMO's and two ABMO's. At the right of each MO, a plot of the wave amplitude appears as a function of internuclear distance. (Recall that the product $\psi\psi^*$ is a measure of electron density at a point; for ease of illustration, we have chosen to plot ψ itself rather than to plot the product $\psi\psi^*$.) Notice that the plots for ψ_1, ψ_2, ψ_3, and ψ_4 contain 0, 1, 2, and 3 *nodes*, respectively; the amplitude function falls to zero at each node, a point of *zero electron density*. Inspection of Fig. A.3 clearly reveals that the energy of the individual MO's *increases* (and their stability correspondingly *decreases*) with increasing number of nodes; this is a general result, reflecting the antibonding situation created by the appearance of these nodes. Electron density is drastically reduced in the region between contiguous atoms, the very region where electron density must be greatest for constructive bonding!

The total pi-energy of butadiene in its ground state is simply determined by introducing the four pi-electrons into the individual orbitals in such a way as to lead to the lowest possible energy configuration. To this end, we apply the Pauli principle, which requires that a maximum of two electrons with their spins paired (mutually *antiparallel*) be introduced into each MO. To minimize the energy of the system, we place the four electrons as indicated in Fig. A.3: two electrons reside in ψ_1 and two in ψ_2. The total pi-energy, E_π, follows as the sum of the individual pi-electron energies; for butadiene, we have

$$E_\pi = 2(\alpha + 1.618\beta) + 2(\alpha + 0.618\beta) = 4\alpha + 4.472\beta$$

Careful inspection of Fig. A.3 reveals that the four MO's of butadiene (two BMO's and two ABMO's) are *symmetrically disposed about the zero (NBMO) level.*

This is a general result for a special class of conjugated systems called *even-alternant systems*, for which energy levels occur in pairs having $\epsilon_j = \alpha \pm m_j\beta$. We define alternant systems as those conjugated pi-systems for which contiguous atoms can be alternately "starred" (*) and "unstarred" (0) such that no two starred or two unstarred positions are ever juxtaposed. An even (odd) alternant system contains an even (odd) number of atoms in mutual conjugation. Some examples of alternant hydrocarbons (AH's) appear in (A.2):

Butadiene	Benzene	Allyl	Benzyl
Even AH	Even AH	Odd AH	Odd AH

(A.2)

Conjugated systems which fail to meet this criterion are termed *nonalternant*; some examples appear in (A.3).

Azulene	Fulvene
Non–AH	Non–AH

(A.3)

Nonalternant hydrocarbons are always encountered in ring systems in which one or more of the rings is *odd*-membered.

The MO's of odd AH's, like those of even AH's, come in pairs having $\epsilon_j = \alpha \pm m_j\beta$. The odd MO occurs as the NBMO ($\epsilon_j = \alpha$). The allyl system serves as an example of an odd AH (Fig. A.4). Electrons present in the NBMO contribute

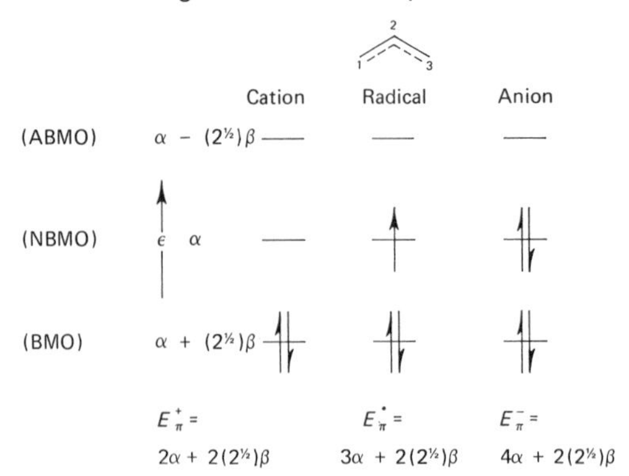

Fig. A.4. *Molecular orbitals of the allyl cation, radical, and anion.*

nothing to the overall stability of the system. They possess the energy of an electron in an isolated $2p$ atomic orbital; their energy is unaffected by delocalization through the pi-system. The total *bonding* energy of the molecule is unaffected by the presence or absence of electrons in its NBMO.

BIBLIOGRAPHY

A. Streitwieser, Jr., "Molecular Orbital Theory for Organic Chemists." Wiley, New York, 1961. For discussion of Hückel MO method, see Chapter 2 and also reference 1, p. 34, both in Streitwieser's book.

M. J. S. Dewar, "The Molecular Orbital Theory of Organic Chemistry." McGraw-Hill, New York, 1969.

A. Liberles, "Introduction to Theoretical Organic Chemistry," Chapter 6. Macmillan, New York, 1968.

A. Liberles, "Introduction to Molecular Orbital Theory." Holt, New York, 1966.

K. Higasi, H. Baba, and A. Rembaum, "Quantum Organic Chemistry." Wiley (Interscience), New York, 1965.

J. D. Roberts, "Notes on Molecular Orbital Calculations." Benjamin, New York, 1961.

Appendix B

THE CONSTRUCTION OF CORRELATION DIAGRAMS

In the construction of correlation diagrams, several steps must be followed. First, it is necessary to ascertain precisely which bonds are broken and which are formed in the reaction being examined. These bonds are then the focal point of the ensuing analysis. Bonds that merely undergo secondary changes, such as rehybridization, are neglected.

Second, one must be concerned with the specific geometry involved in the process at hand. In some cases, that geometry will be dictated by the presence of a rigid bi- or polycyclic system; in others, there is, *a priori*, considerable freedom. For example, cycloaddition reactions could occur through an infinite number of different geometries, varying from approach in parallel planes to approach in perpendicular planes. They are usually visualized in the former manner, since overlap of orbitals is generally thought to be most favorable in that instance. However, other geometries could be analyzed as well, provided that they possess nontrivial symmetry elements (see below).

Third, one examines the system and reduces it so that it possesses the highest number of symmetry elements. Appropriate measures include replacement of unsymmetrically placed alkyl groups by hydrogen atoms and, in some instances, replacement of a heteroatom by a carbon atom with which it is isoelectronic (for example, a neutral boron atom would be replaced by a positively charged carbon atom).

Fourth, one selects the symmetry elements that exist throughout the reaction's course, making sure that "trivial" symmetry elements are not chosen. A trivial symmetry element is one that does not bisect any of the bonds formed or broken in the reaction or which has all MO's either symmetric or antisymmetric with respect to it.

Finally one establishes energy levels with corresponding MO's for the system, classifies them according to their behavior upon operation of the symmetry elements present, and connects levels of like symmetry without crossing levels of like symmetry.

As an example, consider the $[{}_{\pi}4_s + {}_{\pi}2_s]$ cycloaddition of propene to butadiene. The reaction involves directly only the four pi-electron system of butadiene and the two pi-electron system of propene, so we focus attention upon those orbitals. The C—H and C—C sigma-bonds in both systems are involved only in secondary changes, so we do not concern ourselves with them. We assume a geometry for the reaction in which the two systems approach in parallel planes and draw the orbitals involved (Fig. B.1).

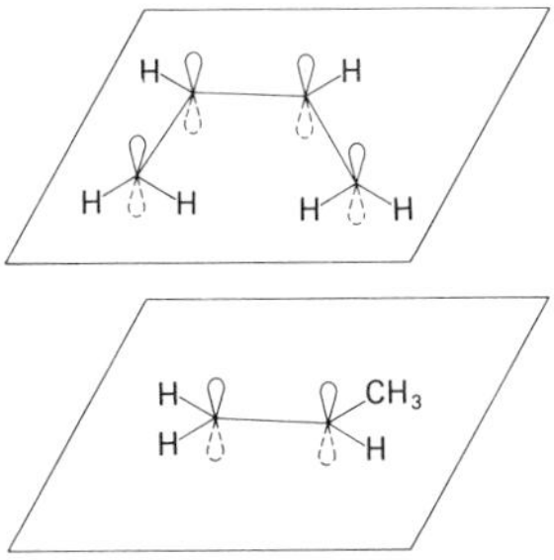

Fig. B.1. *Approach of butadiene and propene in the* $[{}_{\pi}4_s + {}_{\pi}2_s]$ *cycloaddition reaction.*

Clearly, the reaction lacks significant symmetry at this stage, so we remove the troublesome methyl group and replace it with a hydrogen atom (see the discussion following Problem II.3 for a justification). The reaction we analyze then is that of butadiene with ethylene to produce cyclohexene [Eq. (B.1)].

(B.1)

The symmetry element, therefore, is the same as that in the Diels-Alder reaction of butadiene and ethylene, a mirror plane, m (see Fig. II.9 and II.10). The reader should verify that further reduction of the system to a higher order of symmetry is not possible.

Next, we must write down the energy levels of the reactants and products and assign MO's to them. In this instance, the reactants will have three bonding and three antibonding pi-levels (two pi-bonds in butadiene and one in ethylene). The product (cyclohexene) should have two bonding sigma-levels, which represent the newly formed sigma-bonds, and one bonding pi-level. The bonding levels will be mirrored by their antibonding partners. The sigma-bonds must be analyzed in this instance since they are new bonds, not simply rehybridized bonds. As discussed in Chapter II, the precise positioning of the levels is not crucial, but it is most important to know how many are bonding and how many are antibonding. Nonetheless, it is customary to place bonding sigma-bonds below bonding pi-bonds since they are typically much more stable. Likewise an antibonding sigma-bond is *less* stable (more unstable) than an antibonding pi-bond.

A requirement of the MO's is that they must be either symmetric or antisymmetric with respect to the symmetry element(s) describing the reaction, in this case the mirror plane, *m*. An examination of the reacting system reveals that both the butadiene and ethylene portions individually possess the same mirror plane as a symmetry element (each is bisected by it). Therefore, we can establish the MO's of the reactants simply as those of butadiene and ethylene. Also, although we do not need to, we can place the energy levels of the reacting pi-systems in appropriate order simply by consulting one of the references at the end of Appendix A; there, we find that the energy of the ethylene pi-bond lies intermediate between the energies of the corresponding butadiene levels. The energy of the product pi-system should approximate that of ethylene. At this stage, we may write down the energy levels for the reaction (Fig. B.2).

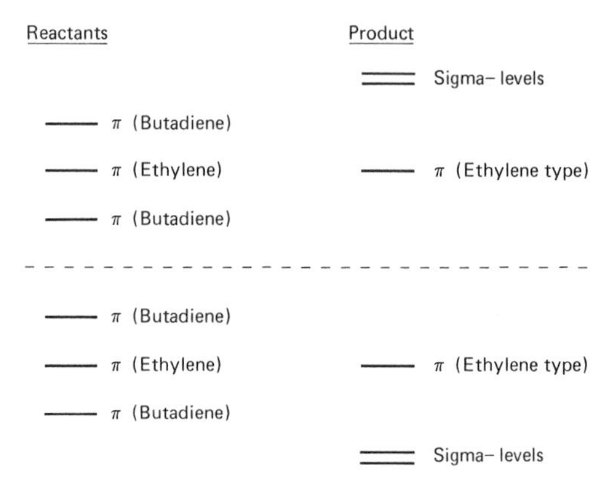

Fig. B.2. *Energy levels for the* $[_{\pi}4_{s} + {}_{\pi}2_{s}]$ *cycloaddition of butadiene and ethylene.*

As previously stated, there is no problem establishing the MO's for reactants. However, it is clear that the individual sigma-bonds in cyclohexene do not meet the requirement of being symmetrical with respect to reflection through the mirror plane, *m*. Accordingly, it is necessary to generate combinations of sigma-bonds, in much the same manner that we generated butadiene pi-bonds from individual *p* orbitals (see Appendix A). That is most easily accomplished by taking simple sums and differences of the two sigma-bonds. The bonding sigma-levels are generated by taking combinations of the bonding sigma-bonds. The antibonding sigma-levels are generated by taking combinations of the antibonding sigma-bonds. If we call the individual bonding sigma-bonds σ_A and σ_B we will generate the bonding levels $(\sigma_A + \sigma_B)$ and $(\sigma_A - \sigma_B)$, as shown in Fig. B.3. Both of these

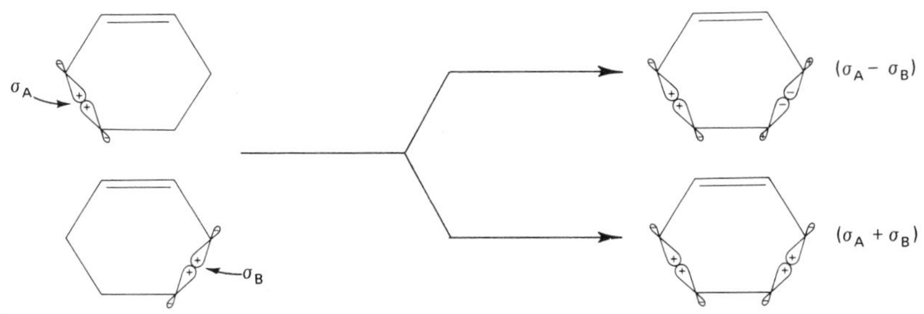

Fig. B.3. *The generation of bonding sigma-levels in cyclohexene.*

levels are bonding, since each sigma-bond is bonding. Furthermore, the combination now satisfies the symmetry requirements of the system. The MO $(\sigma_A + \sigma_B)$ is symmetric to reflection through *m*, and the MO $(\sigma_A - \sigma_B)$ is antisymmetric to the operation. The two levels should vary slightly in energy due to secondary interactions (shown in red) between the sigma-bonds (Fig. B.4). Since the interactions are favorable for the combination $(\sigma_A + \sigma_B)$ but unfavorable for the combination $(\sigma_A - \sigma_B)$, the former should be somewhat more stable.

In similar manner, we generate the antibonding sigma-levels from the individual antibonding sigma-bonds, σ_C^* and σ_D^* (Fig. B.5). Of these two antibonding combinations, $(\sigma_C^* + \sigma_D^*)$ should be the least unstable, due to favorable secondary interactions.

As the final step in construction of the correlation diagram, we place the MO's we have generated beside their appropriate energy levels, and classify them according to their symmetry with respect to the symmetry operation, in this case reflection through plane *m* (Fig. B.6).

Fig. B.4. *Secondary interactions between orbitals of bonding sigma-levels.*

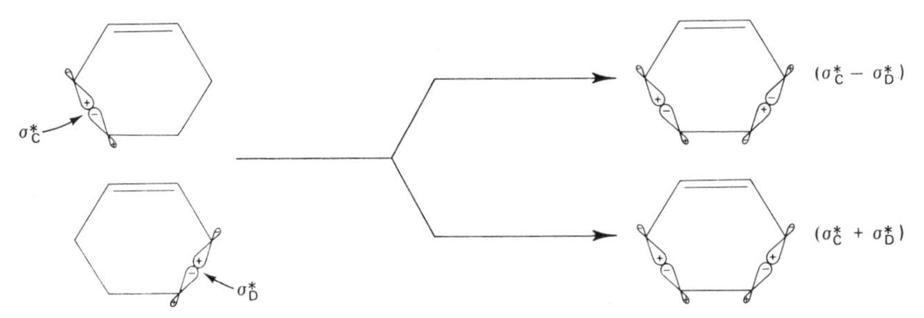

Fig. B.5. *The generation of antibonding sigma-levels in cyclohexene.*

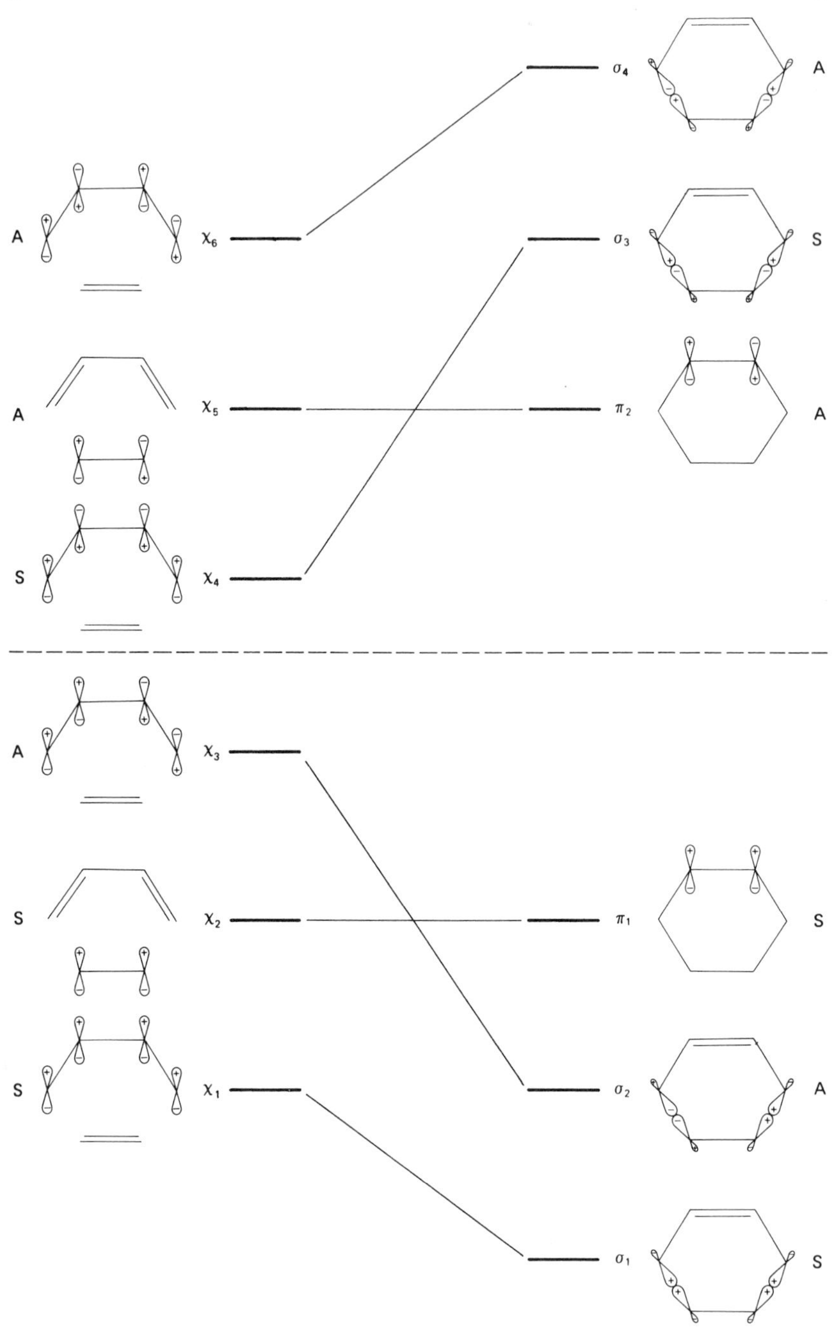

Fig. B.6. *Correlation diagram for the* $[{}_{\pi}4_s + {}_{\pi}2_s]$ *cycloaddition of butadiene and ethylene.*

The reader is reminded that each of these is a molecular orbital and involves all six atoms of the system. That no orbital lobes are drawn on the ethylene portion of the orbital X_1, for example, is intended to indicate only that the coefficients of the p orbitals at those atoms are zero. Similar considerations apply to atoms in other MO's wherein the orbital lobes are not drawn. In order to complete the construction of the correlation diagram, energy levels of like symmetry are connected in such a way that no two levels of like symmetry cross. This is a statement of the familiar "noncrossing rule." The actual application of correlation diagrams to the analysis of concerted reactions is discussed in Chapter II.

As a second example, we choose the $[_{\sigma}2_s + _{\sigma}2_s]$ exchange reaction between hexadeuterioethane and ethane to produce 1,1,1-trideuterioethane. We make the assumption that the approach of the C—C sigma-bonds is parallel [Eq. (B.2)].

$$\begin{matrix} D_3C-CD_3 \\ H_3C-CH_3 \end{matrix} \longrightarrow \begin{matrix} D_3C & CD_3 \\ | & | \\ H_3C & CH_3 \end{matrix} \qquad \text{(B.2)}$$

Further, in order to achieve the highest symmetry possible, we replace the deuterium atoms in hexadeuterioethane with hydrogen atoms. Next, we draw out the orbitals of the bonds involved (Fig. B.7). It is evident that there are two symmetry planes, m_1 and m_2. The energy-level diagram should have two bonding and two antibonding sigma-levels for reactants and two bonding and two antibonding sigma-levels for the products (Fig. B.8).

It is now our task to generate molecular orbitals for each of the energy levels. The molecular orbitals must be either symmetric or antisymmetric to the operation of each of the symmetry elements. Once again, it is clear that the individual sigma-bonds present in each of the reactants do not satisfy the symmetry requirement. Each is individually symmetric with respect to m_1, but not to m_2. We therefore need to generate the MO's. This is again accomplished by taking sums and

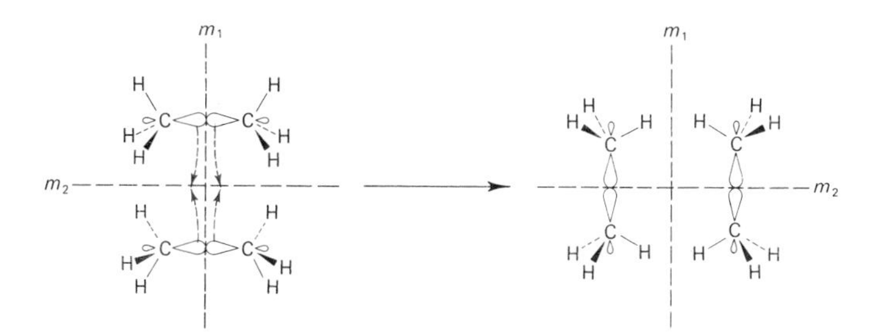

Fig. B.7. *Orbitals directly involved in the $[_{\sigma}2_s + _{\sigma}2_s]$ exchange reaction of two ethane molecules.*

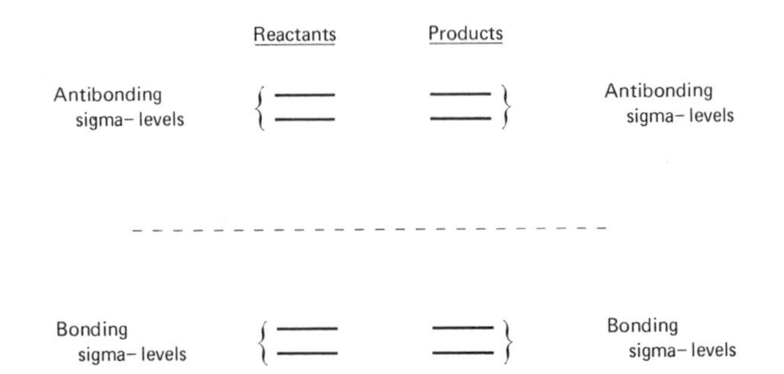

Fig. B.8. *Energy-level diagram for the* $[_{\sigma}2_s + {}_{\sigma}2_s]$ *exchange reaction of two ethane molecules.*

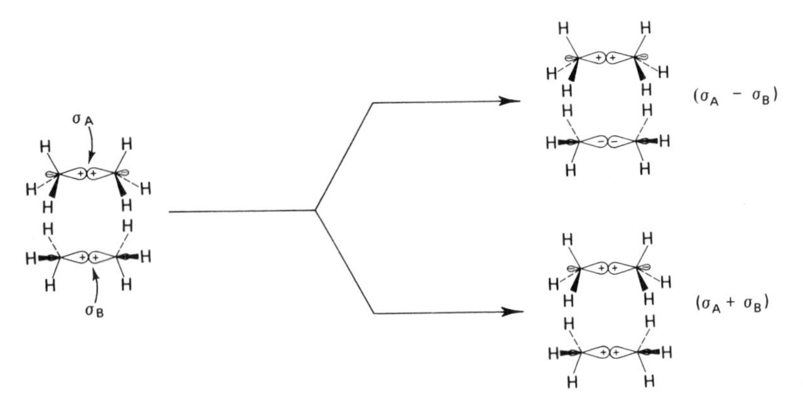

Fig. B.9. *Generation of bonding sigma-levels for the reactants in the* $[_{\sigma}2_s + {}_{\sigma}2_s]$ *exchange reaction of two ethane molecules.*

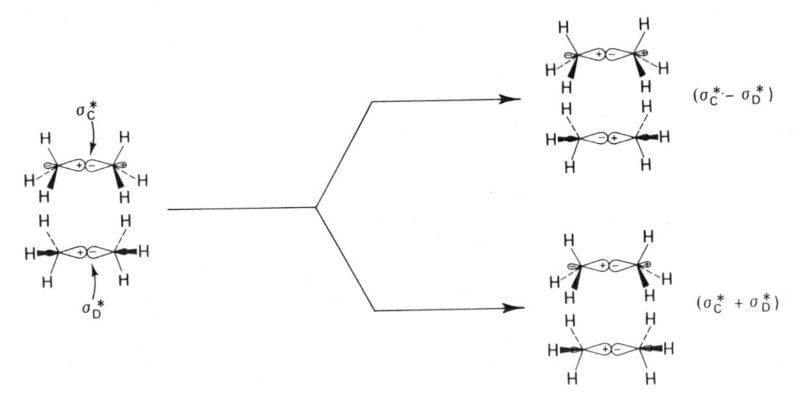

Fig. B.10. *Generation of antibonding sigma-levels for the reactants in the* $[_{\sigma}2_s + {}_{\sigma}2_s]$ *exchange reaction of two ethane molecules.*

differences of the individual sigma-bonds. The bonding levels are generated from the bonding sigma-bonds (Fig. B.9).

The antibonding levels are generated from the individual antibonding sigma-bonds (Fig. B.10).

In precisely analogous manner, it is possible to generate the MO's of the product system. The reader may verify this. The reactant and product MO's are matched with their energy levels in the correlation diagram. The symmetries of the orbitals are noted, and levels of identical symmetry are connected (Fig. B.11).

Note that it is very important to classify each MO carefully and to avoid confusion in the ordering of symmetry labels. An SA molecular orbital correlates with another SA molecular orbital, not an AS molecular orbital. Further, note

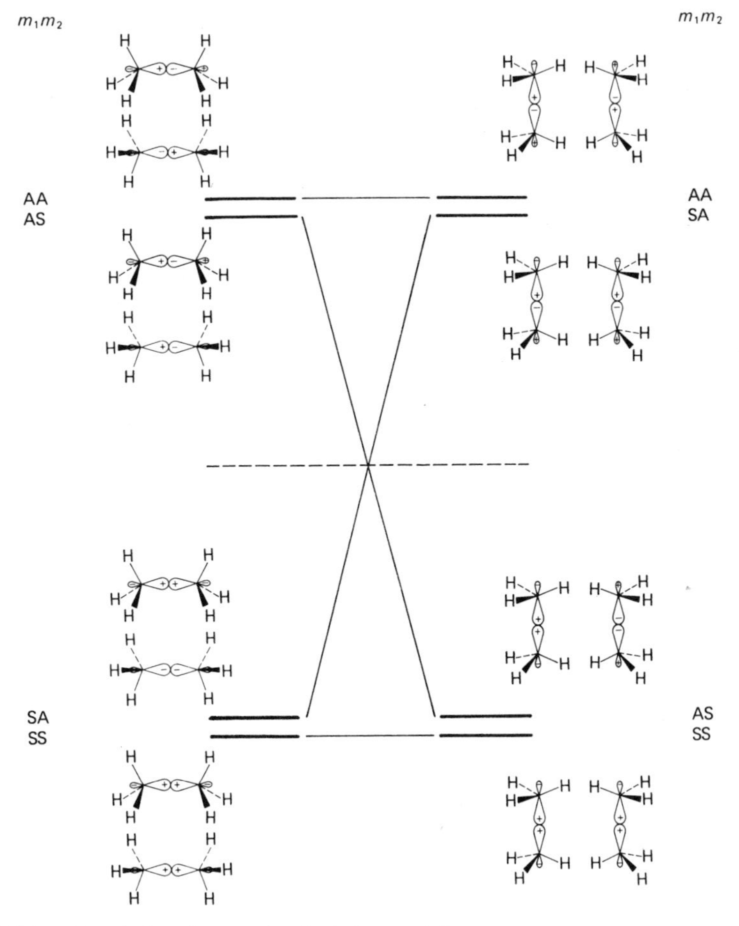

Fig. B.11. *Correlation diagram for the $[_{\sigma}2_s + {}_{\sigma}2_s]$ exchange reaction of two ethane molecules.*

that there is a C_2 axis of symmetry passing between the two ethane molecules (Fig. B.12). We have not included it in the analysis because it is a "trivial" symmetry element that does not bisect any bonds.

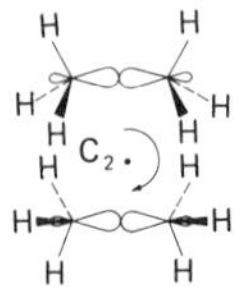

Fig. B.12. *The trivial C_2 symmetry element in the $[_{\sigma}2_s + {}_{\sigma}2_s]$ exchange reaction of two ethane molecules.*

Part III

ANSWERS TO PROBLEMS

ANSWERS TO PROBLEMS IN CHAPTER I

PROBLEM I.1

(a) Disrotatory.
(b) Conrotatory.
(c) Conrotatory.

PROBLEM I.2

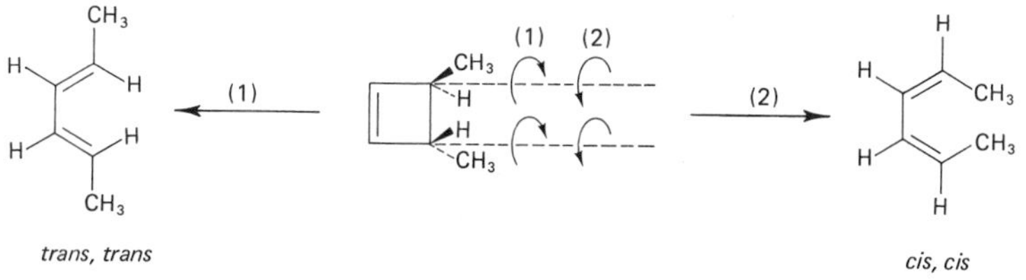

trans, trans *cis, cis*

Conrotatory process 1 leads to *trans, trans* diene, whereas conrotatory process 2 leads to *cis, cis* diene. In process 2, molecular motions will lead to increasing interaction of the methyl groups as the transition state is approached. This unfavorable interaction is absent in process 1, so process 1 should be favored.

PROBLEM I.3

Eq. (I.2). *Reference*: E. Vogel, W. Grimme, and E. Dinne, *Tetrahedron Lett.* p. 391 (1965).

Thermal (a $4q + 2$ system, disrotatory).

Eq. (I.3). *Reference*: R. B. Bates and D. A. McCombs, *Tetrahedron Lett.* p. 977 (1969).

Thermal (a $4q + 2$ system, disrotatory).

Eq. (I.4). *Reference*: E. Vogel, *Liebigs Ann. Chem.* **615**, 14 (1958).

Thermal (a $4q$ system, conrotatory).

Problem I.1

(a) *Reference*: W. G. Dauben, R. G. Cargill, R. M. Coates, and J. Saltiel, *J. Amer. Chem. Soc.* **88**, 2742 (1966).

Photochemical (a $4q$ system, disrotatory).

(b) *Reference*: R. Huisgen, A. Dahmen, and H. Huber, *J. Amer. Chem. Soc.* **89**, 7130 (1967).

Thermal (a $4q$ system, conrotatory).

(c) *Reference*: W. D. Huntsman and H. J. Wristers, *J. Amer. Chem. Soc.* **89**, 342 (1967).

Thermal (a $4q$ system, conrotatory).

PROBLEM I.4

(a) *Reference*: W. R. Roth and J. König, *Liebigs Ann. Chem.* **699**, 24 (1966).

[1,5].

(b) *Reference*: J. W. Baldwin and C. H. Armstrong, *J. Chem. Soc.* (D), 631 (1970).

[2,3]. The above authors discuss [2,3] sigmatropic shifts in this and similar systems. In this instance, radical dissociation–recombination competes.

(c) *Reference*: Gy. Frater and H. Schmid, *Helv. Chim. Acta* **51** 190 (1968).

[5,5]. The structure shown is implicated as an intermediate.

PROBLEM I.5

Reference: W. R. Roth, J. König, and K. Stein, *Chem. Ber.* **103** 426 (1970).

Each is formed by a suprafacial [1,5] hydrogen shift.

H
C_2H_5 CH_3 D CH_3 → H D C_2H_5 CH_3 CH_3 ≡ H CH_3 D C_2H_5 CH_3

H_3C C_2H_5 D CH_3 H → D CH_3 H_3C C_2H_5 H ≡ D CH_3 C_2H_5 H_3C H

PROBLEM I.6

(a) Suprafacial with retention:

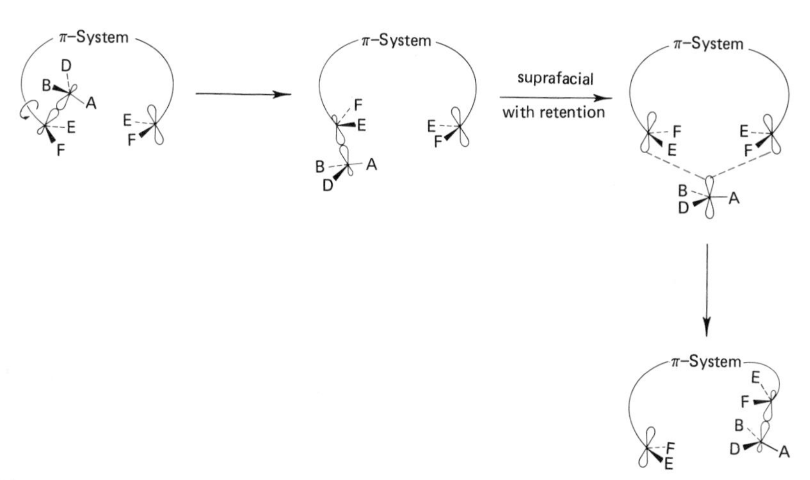

(b) Antarafacial with inversion:

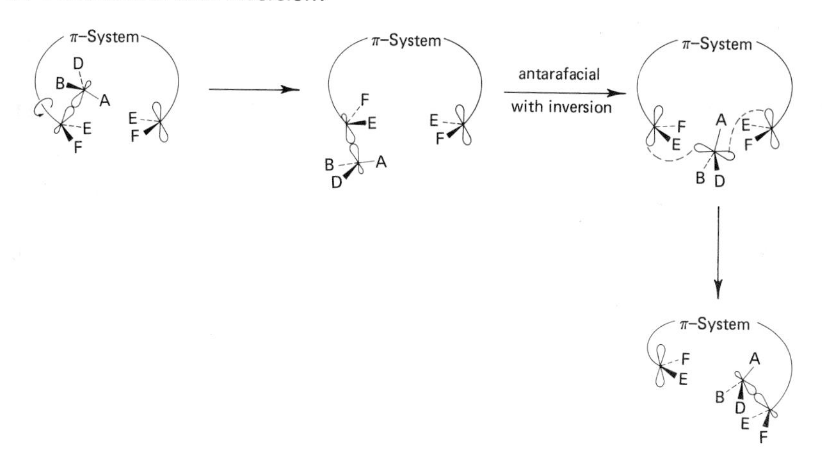

PROBLEM I.7

Reference: J. L. M. A. Schlatmann, J. Pot, and E. Havinga, *Rec. Trav. Chim.* **83**, 1173 (1964).

[1,7], so $1 + j = 4n$. If thermally induced and concerted, the migration should be antarafacial.

PROBLEM I.8

Reference: R. B. Woodward and T. J. Katz, *Tetrahedron* **5**, 70 (1959).

[3,3] Sigmatropic shift, suprafacial on each component. Since $1 + j = 4n + 2$, the reaction should be thermally allowed.

PROBLEM I.9

Reference: R. C. Cookson, *Chem. Brit.* **5**, 6 (1969).

One can envision two processes that would give the skeletal arrangement of the product, a [1,3] and a [3,3] sigmatropic shift:

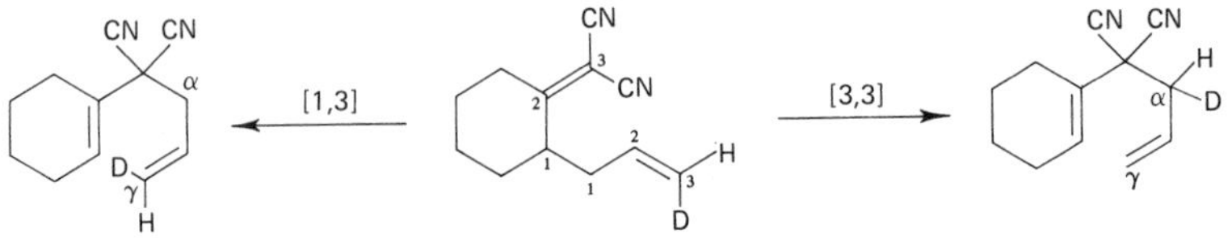

The [1,3] shift leads to the presence of deuterium in position γ. The [3,3] shift leads to the presence of deuterium in position α. The reaction is photochemical and must be suprafacial because of the system involved. An examination of Table I.3 reveals that $i + j = 4n$ is required for the photochemical reaction. Thus, a [1,3] sigmatropic shift should be observed and deuterium should be found at position γ. This prediction has been confirmed by the experimental results of Cookson.

PROBLEM I.10

(a) This is a thermal [1,4] suprafacial shift in a carbonium ion. Table I.4 indicates that the change is forbidden to proceed with retention at the migrating group ($1 + 4 = 5 = 4n + 1$). Since the rules reverse if the migrating group undergoes inversion, we expect *inversion* at the migrating group.

(b) Similar examination of Table I.4 indicates that the [1,4] shift in the anion should proceed with *retention*.

PROBLEM I.11

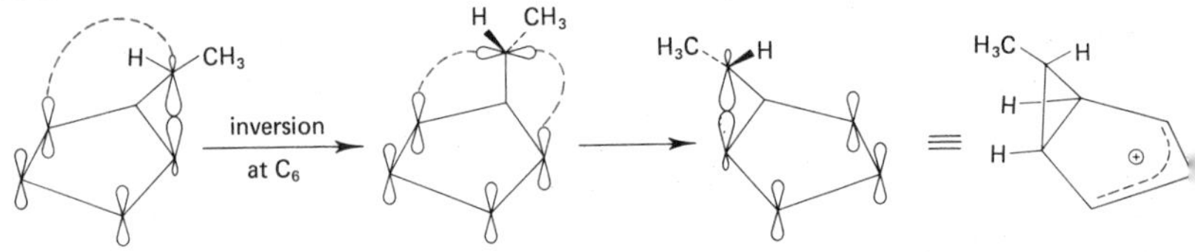

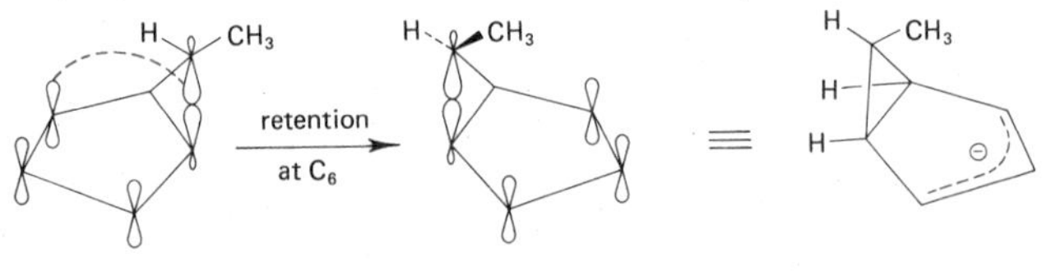

PROBLEM I.12

(a) [4 + 4].
(b) [4 + 2].
(c) [4 + 2]. (This is termed a cycloreversion reaction.)

PROBLEM I.13

(a)

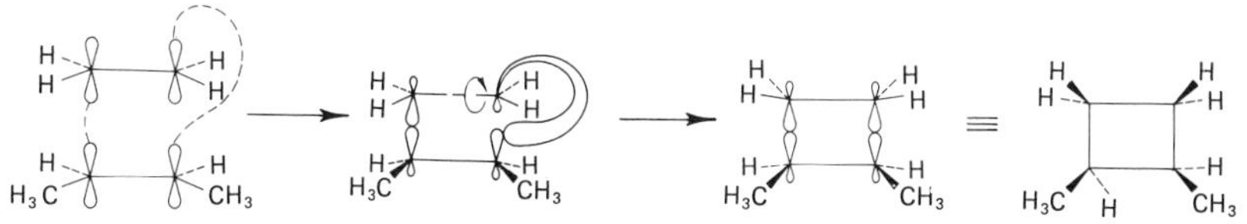

cis-1, 2-Dimethylcyclobutane

(b)

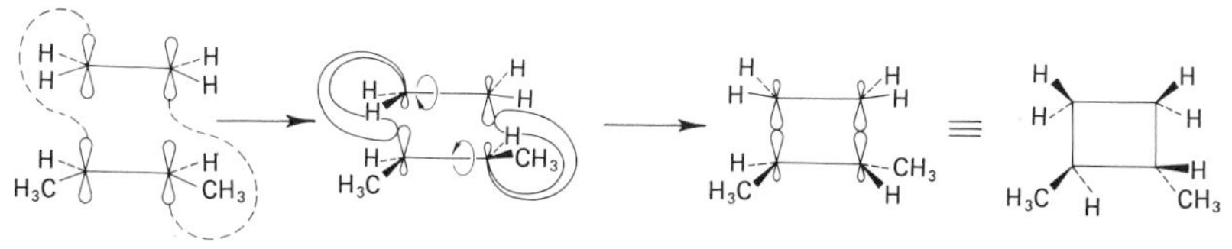

trans-1, 2-Dimethylcyclobutane

PROBLEM I.14

(a) (A) (supra)-(supra)

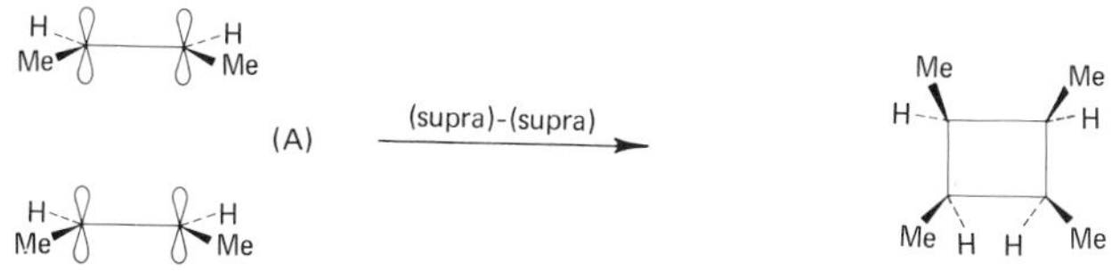

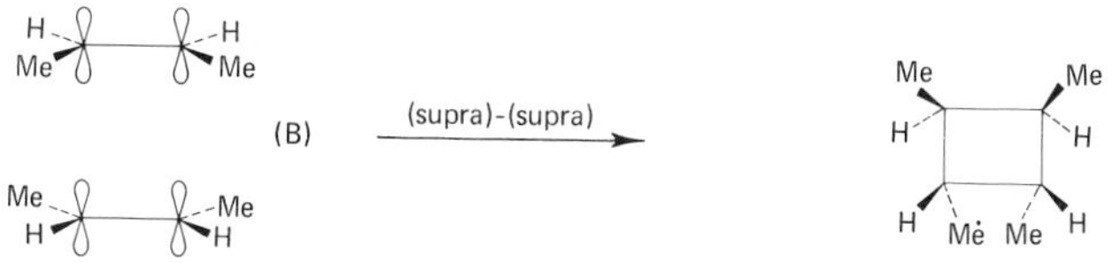

(B) (supra)-(supra)

(b) (A) or (B) (supra)-(antara)

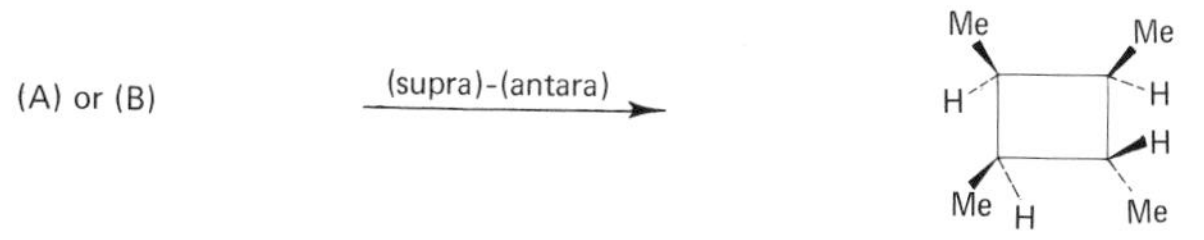

(c) (A) or (B) (antara)-(antara)

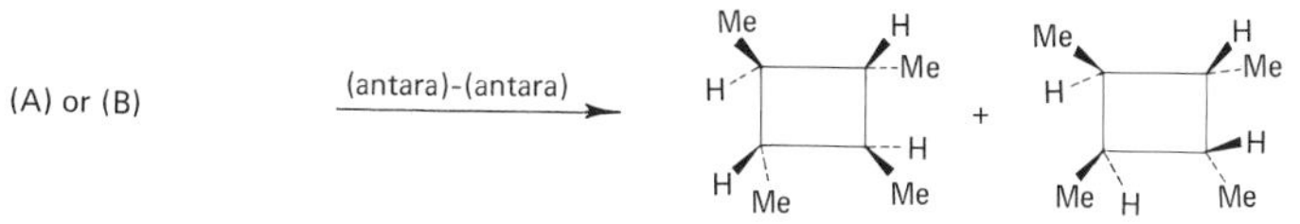

PROBLEM I.15

(a) *Reference*: K. N. Houk and C. R. Watts, *Tetrahedron Lett.* p. 4025 (1970).
[6 + 4]. Suprafacial on tropone. No stereochemical probe at other component.

(b) *Reference*: K. Kraft and G. Koltzenburg, *Tetrahedron Lett.* pp. 4357 and 4723 (1967).
[2 + 2]. Suprafacial on one component, antarafacial on the other.

(c) *Reference*: J. Saltiel and L.-S. Ng Lim, *J. Amer. Chem. Soc.* **91**, 5404 (1969).
[2 + 2]. Suprafacial on double bond. No stereochemical probe possible at the triple bond.

PROBLEM I.16

Problem I.12

(a) *Reference*: E. Vogel, W. Grimme, W. Meckel, H. J. Riekel, and J. F. M. Oth, *Angew. Chem. Int. Ed. Engl.* **5**, 590 (1966).
$[4_s + 4_s]$, $m + n = 8 = 4q$. Photochemically allowed. The structure shown is implicated as an intermediate.

(b) *Reference*: H. M. R. Hoffmann, D. R. Joy, and A. K. Suter, *J. Chem. Soc.* (B), 57 (1968).
$[4_s + 2_s]$. $m + n = 6 = 4q + 2$. Thermally allowed.

(c) *Reference*: J. N. Hines, M. J. Peagram, G. H. Whitham, and M. Wright, *Chem. Commun.* p. 1593 (1968).
$[4_s + 2_s]$, $m + n = 6 = 4q + 2$. Thermally allowed. The above authors prepared *trans*-cyclooctene by this route.

Problem I.15

(a) $[6_s + 4_s]$, $m + n = 10 = 4q + 2$. Thermally allowed.
(b) $[2_s + 2_a]$, $m + n = 4 = 4q$. Thermally allowed.
(c) $[2_s + 2_s]$, $m + n = 4 = 4q$. Photochemically allowed.

PROBLEM I.17

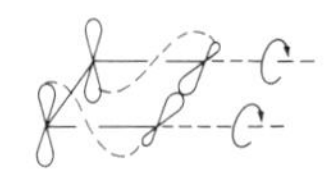

PROBLEM I.18

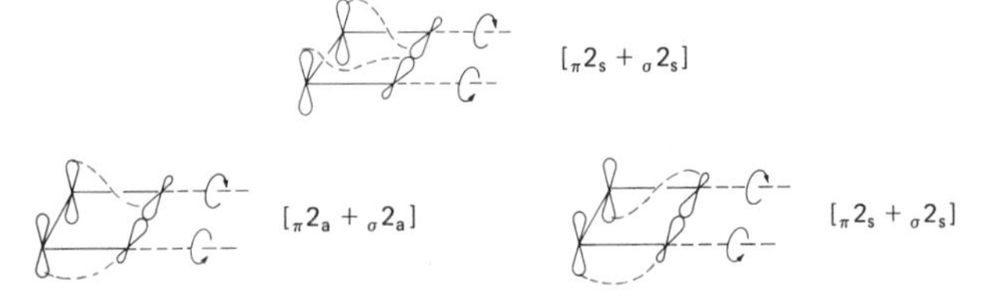

This is a $4q$ system and the disrotatory process may be classified as either $[_{\pi}2_a + {}_{\sigma}2_a]$ or $[_{\pi}2_s + {}_{\sigma}2_s]$. According to Table I.5, this should be a photochemically allowed–thermally forbidden process. That is the same conclusion reached from examination of Table I.1.

PROBLEM I.19

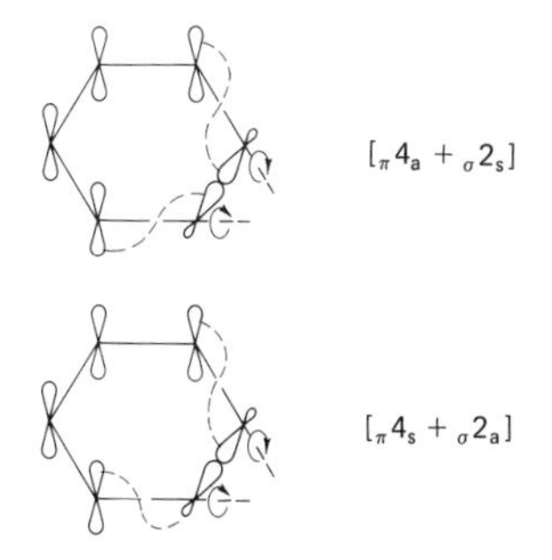

(Another $[_{\pi}4_a + {}_{\sigma}2_s]$ classification could be presented.) This is a $4q + 2$ system and addition is suprafacial on one component and antarafacial on the other. The process is predicted (Table I.5) to be allowed photochemically and forbidden thermally. The same conclusion is reached from examination of Table I.1.

PROBLEM I.20

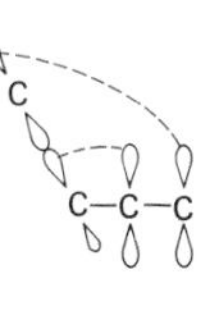

$[_{\sigma}2_a + {}_{\pi}2_s]$

or

C

C—C—C

$[_{\sigma}2_s + {}_{\pi}2_a]$

Allowed thermally.

PROBLEM I.21

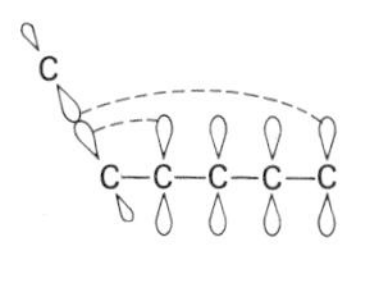

$[_{\pi}4_s + {}_{\sigma}2_s]$

or

C

C—C—C—C—C

$[_{\pi}4_a + {}_{\sigma}2_a]$

Allowed thermally.

PROBLEM I.22

Problem I.8

The reaction in Problem I.8 is $[_{\sigma}2_s + {}_{\pi}2_s + {}_{\pi}2_s]$ or $[_{\sigma}2_a + {}_{\pi}2_s + {}_{\pi}2_a]$ or $[_{\sigma}2_s + {}_{\pi}2_a + {}_{\pi}2_a]$. Of course, all classifications lead to the same conclusion: thermally allowed, photochemically forbidden.

Problem I.1

(a) $[_{\sigma}2_s + {}_{\pi}2_s]$ or $[_{\sigma}2_a + {}_{\pi}2_a]$. Thermally forbidden, photochemically allowed.
(b) $[_{\pi}6_a + {}_{\sigma}2_s]$ or $[_{\pi}6_s + {}_{\sigma}2_a]$. Thermally allowed, photochemically forbidden.
(c) $[_{\sigma}2_s + {}_{\pi}2_a]$ or $[_{\sigma}2_a + {}_{\pi}2_s]$. Thermally allowed, photochemically forbidden.

PROBLEM I.23

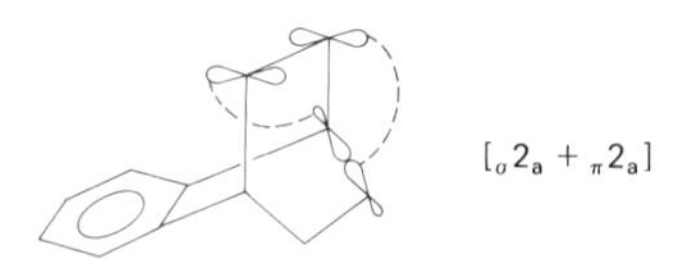

$[_{\sigma}2_a + {}_{\pi}2_a]$

Allowed photochemically.

PROBLEM I.24

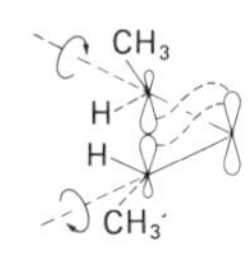

The reaction involves addition of a sigma-bond to a *p* orbital. However, how does one classify a *p* orbital? Woodward and Hoffmann have suggested ω as a symbol. When addition occurs to the same lobe of the *p* orbital, the addition is suprafacial. When addition occurs to opposite lobes of the *p* orbital, the addition is antarafacial. In this case, there are no electrons in the orbital, so, if we consider the interactions shown above, addition to the *p* orbital is classified as $_{\omega}0_s$, and the reaction is classified $[_{\sigma}2_s + {}_{\omega}0_s]$. Note that the reaction could also be classified as $[_{\sigma}2_a + {}_{\omega}0_a]$. Allowed thermally.

ANSWERS TO PROBLEMS IN CHAPTER II

PROBLEM II.1

The ground state of cyclobutene is $\sigma^2\pi^2$. It correlates with $\psi_1^2\psi_3^2$, a doubly excited state of butadiene.

PROBLEM II.2

For butadiene, $\psi_1^2\psi_2^1\psi_3^1$. For cyclobutene, $\sigma^2\pi^1\pi^{*1}$. They correlate only in the black pathway. In the red pathway $\psi_1^2\psi_2^1\psi_3^1 \longleftrightarrow \sigma^1\pi^2\sigma^{*1}$ and $\sigma^2\pi^1\pi^{*1} \longleftrightarrow \psi_1^1\psi_2^2\psi_4^1$.

PROBLEM II.3

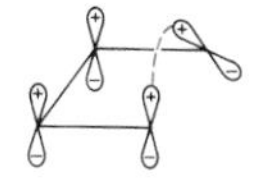

ψ_1, Bonding

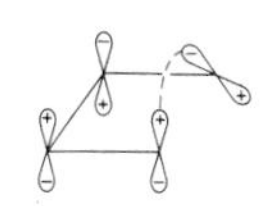

ψ_2, Antibonding

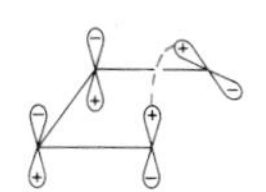

ψ_3, Bonding

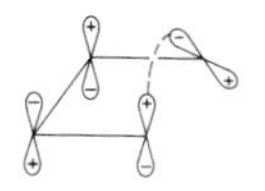

ψ_4, Antibonding

PROBLEM II.4

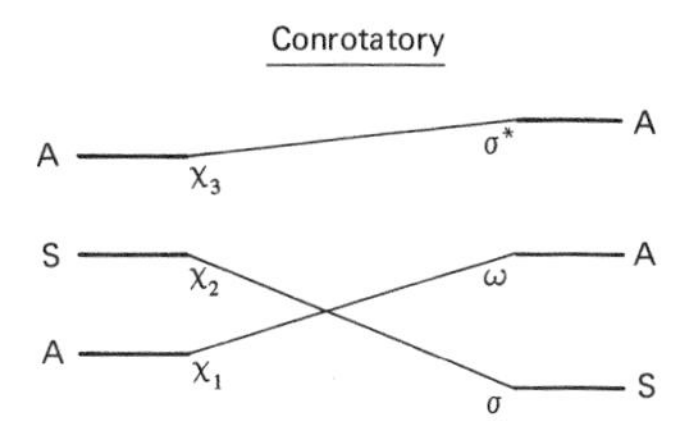

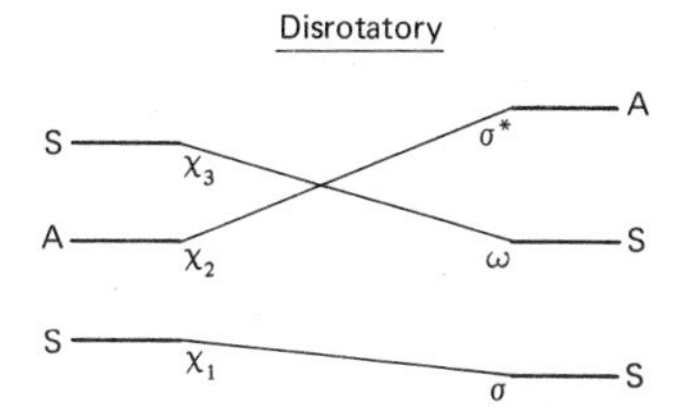

PROBLEM II.5

Two electrons are involved in the carbonium ion process, so thermally,

Conrotatory	Disrotatory
$\chi_1^2 \longleftrightarrow \omega^2$	$\chi_1^2 \longleftrightarrow \sigma^2$
Thermally forbidden	Thermally allowed

For the first excited states,

Conrotatory	Disrotatory
$\chi_1^1\chi_2^1 \longleftrightarrow \sigma^1\omega^1$	$\chi_1^1\chi_2^1 \longleftrightarrow \sigma^1\sigma^{*1}$
Photochemically allowed	Photochemically forbidden

PROBLEM II.6

Four electrons are involved in the anion interconversion, so thermally,

Conrotatory	Disrotatory
$\chi_1^2\chi_2^2 \longleftrightarrow \sigma^2\omega^2$	$\chi_1^2\chi_2^2 \longleftrightarrow \sigma^2\sigma^{*2}$
Thermally allowed	Thermally forbidden

PROBLEM II.7

Reference: H. C. Longuet-Higgins and E. W. Abrahamson, *J. Amer. Chem. Soc.* **87**, 2045 (1965).

Allyl radical ground state: $\chi_1^2\chi_2^1$

Cyclopropyl radical ground state: $\sigma^2\omega^1$

Neither process interconverts them! See the above authors for a discussion of possible effects of configuration interaction in determining the course of the reaction. Note, in any event, that the cyclopropyl radical ⇌ allyl radical interconversion should confront a symmetry-imposed energy barrier regardless of mode of transformation.

PROBLEM II.8

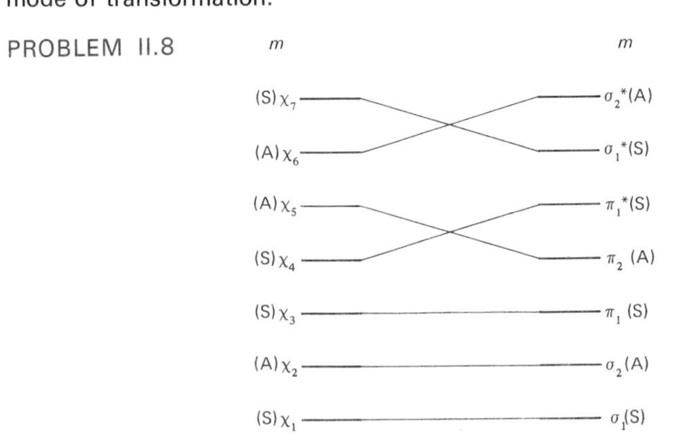

PROBLEM II.9

(a) Eight electrons are involved in the thermal pentadienyl anion + ethylene addition, and $X_1^2X_2^2X_3^2X_4^2 \longleftrightarrow \sigma_1^2\sigma_2^2\pi_1^2\pi_1^{*2}$—the reaction is forbidden.

(b) Six electrons are involved in the pentadienyl cation + ethylene addition, so $X_1^2X_2^2X_3^2 \longleftrightarrow \sigma_1^2\sigma_2^2\pi_1^2$. In this case, ground states *do* correlate, so the reaction is thermally allowed.

PROBLEM II.10

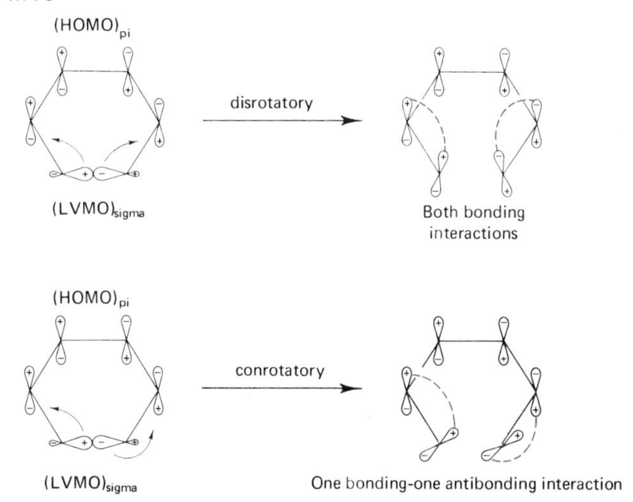

The disrotatory process should be favored.

PROBLEM II.11

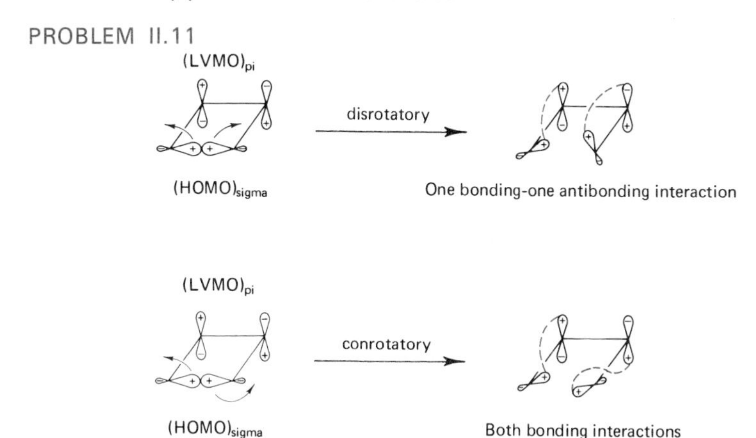

Conrotatory transformation should be favored. The same prediction is obtained by examining the $(\mathrm{LVMO})_{\mathrm{sigma}}$–$(\mathrm{HOMO})_{\mathrm{pi}}$ interaction.

PROBLEM II.12

In the thermal reaction, $(\text{HOMO})_{\text{butadiene}}$ is ψ_2:

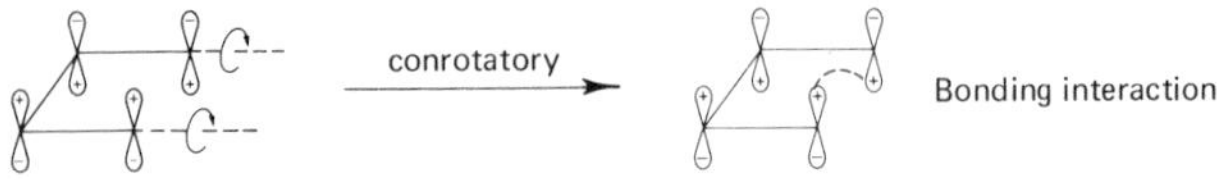

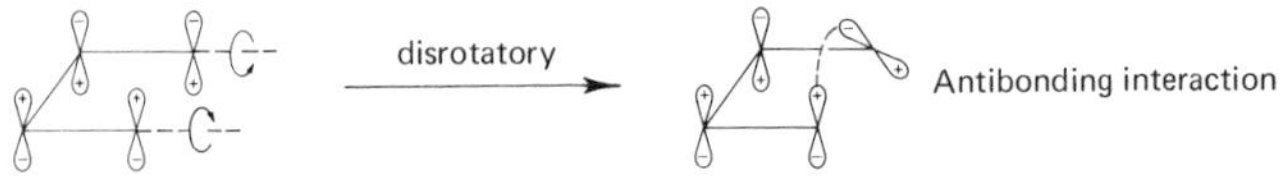

conrotatory → Bonding interaction

disrotatory → Antibonding interaction

In the conrotatory transformation, the terminal orbitals interact in a bonding manner, whereas in the disrotatory transformation, they interact in an antibonding manner. The reaction should therefore proceed in the conrotatory manner.

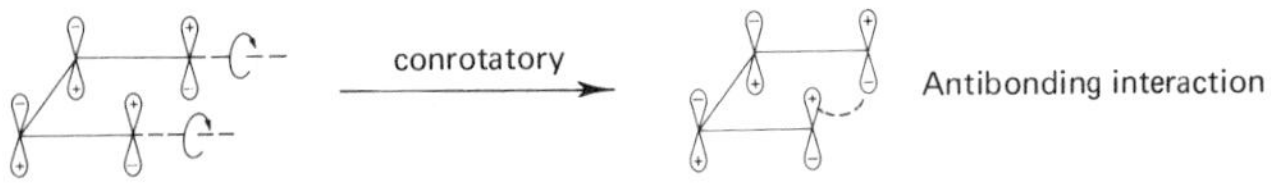

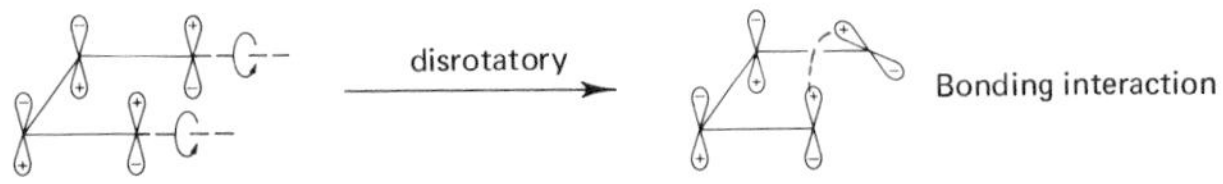

conrotatory → Antibonding interaction

disrotatory → Bonding interaction

The disrotatory process is predicted to be preferred.

PROBLEM II.13

(a) $(\text{HOMO})_{\text{pentadienyl cation}}$:

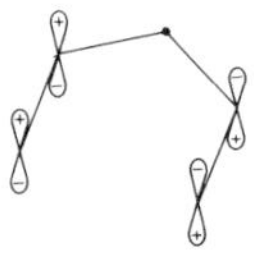

Only in the *conrotatory* process do the terminal lobes interact in the bonding manner. Conrotatory closure is predicted.

(b) $(\text{HOMO})_{\text{pentadienyl anion}}$:

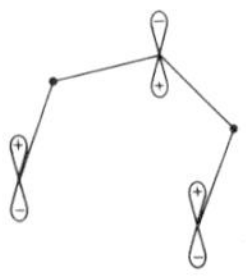

Only in the *disrotatory* process do the terminal orbitals interact in the bonding manner. Disrotatory closure is predicted.

PROBLEM II.14

References: K. N. Houk and R. B. Woodward, *J. Amer. Chem. Soc.* **92**, 4143, 4145 (1970); K. N. Houk, *Tetrahedron Lett.* p. 2621 (1970) and references cited therein for additional reactions.

We examine the *endo* transition state:

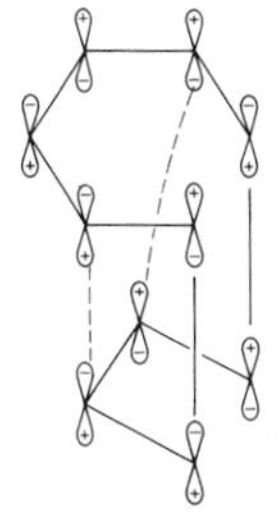

$(LVMO)_{hexatriene}$

$(HOMO)_{butadiene}$

The secondary interactions are found to be *unfavorable*, so the *exo* transition state should be favored. This prediction has received verification in the reactions of cycloheptatriene and tropone with 2,5-dimethyl-3,4-diphenylcyclopentadienone.

PROBLEM II.15

The hydrogen may be passed from top to bottom of the pi-system with maintenance of bonding interactions.

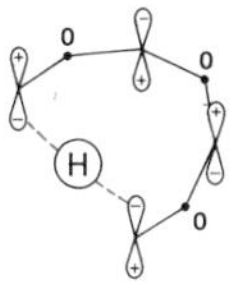

PROBLEM II.16

The transition state is

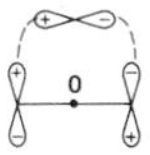

Thus, if inversion occurs at the migrating carbon atom, the [1,3] suprafacial shift is allowed.

PROBLEM II.17

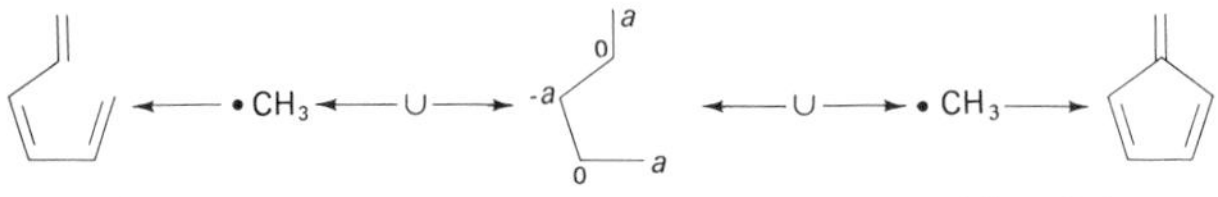

$\Delta E_\pi = 2\beta(1 \cdot a) = 2a\beta$

$\Delta E_\pi = 2\beta(1 \cdot a + 1 \cdot 0) = 2a\beta$

Since the ΔE_π's for formation of fulvene and its acyclic analog, 1,3,5-hexatriene, are the same, fulvene is nonaromatic.

Note: The reader may have noted that there is more than one way to form fulvene by union; for example, the scheme below yields fulvene by union of two allyl units:

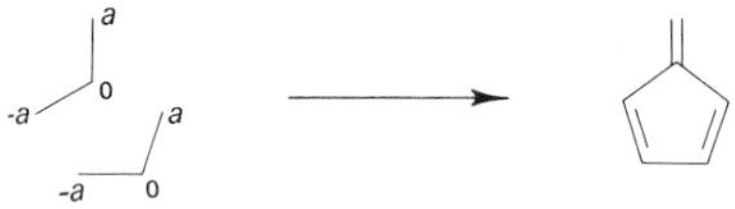

$\Delta E_\pi = 2\beta\,[0 \cdot a + (-a) \cdot (-a)] = 2a^2\beta$

The ΔE_π obtained differs from that which results from union of pentadienyl with methyl. This is easily seen by calculating the a's and determining ΔE_π's in terms of β. For pentadienyl, normalization leads to $a^2 + (-a)^2 + a^2 = 1; a = \pm 1/\sqrt{3}$. For allyl, $a^2 + (-a)^2 = 1; a^2 = \frac{1}{2}$. The ΔE_π's can then be determined:

Pentadienyl + methyl	Allyl + allyl
$\Delta E_\pi = 2a\beta = 2/\sqrt{3}\beta = 1.15\beta$	$\Delta E_\pi = 2a^2\beta = 1.0\beta$

The union of pentadienyl with methyl results in a greater ΔE_π than union of allyl with allyl. When using the PMO method, the reader should always choose the approach that yields the greatest ΔE_π.

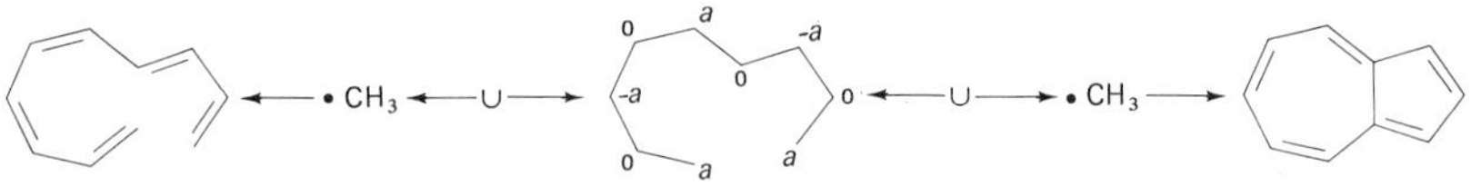

$\Delta E_\pi = 2\beta(1 \cdot a) = 2a\beta$

$\Delta E_\pi = 2\beta(1 \cdot a + 1 \cdot a + 1 \cdot 0) = 4a\beta$

Azulene is predicted to be aromatic (as it is!)

PROBLEM II.18

The transition state for this interconversion is isoconjugate with benzene, which is aromatic if Hückel, but antiaromatic if anti-Hückel (Möbius). The transition state for the disrotatory transformation is Hückel, whereas the transition state for the conrotatory transformation is anti-Hückel. Thus, the disrotatory process should be observed.

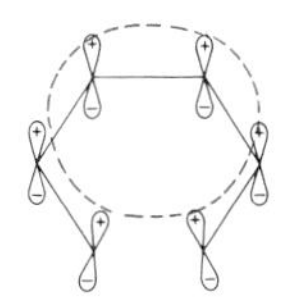

Disrotatory transformation, Hückel transition state, isoconjugate with benzene—*aromatic*

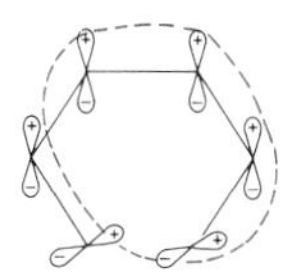

Conrotatory transformation, anti-Hückel transition state, isoconjugate with benzene—*antiaromatic*

PROBLEM II.19

We examine the transition state in each instance:

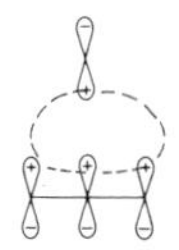

[1,3] Carbon migration with retention, Hückel transition state, isoconjugate with cyclobutadiene, a $4n$ system—*antiaromatic*

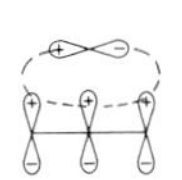

[1,3] Carbon migration with inversion, anti-Hückel transition state, isoconjugate with cyclobutadiene, a $4n$ system—*aromatic*

The migration with inversion is allowed.

PROBLEM II.20

Reference: W. von E. Doering and W. R. Roth, *Tetrahedron* **18**, 67 (1962).
The transition states are

(1)

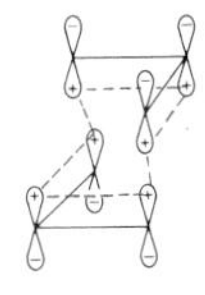

Huckel transition state, isoconjugate with benzene, a Huckel aromatic system

(2)

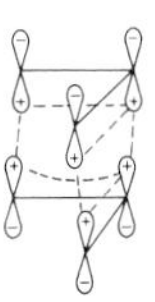

Huckel transition state, isoconjugate with bicyclohexatriene,

The latter system is aromatic, but less so than benzene, since

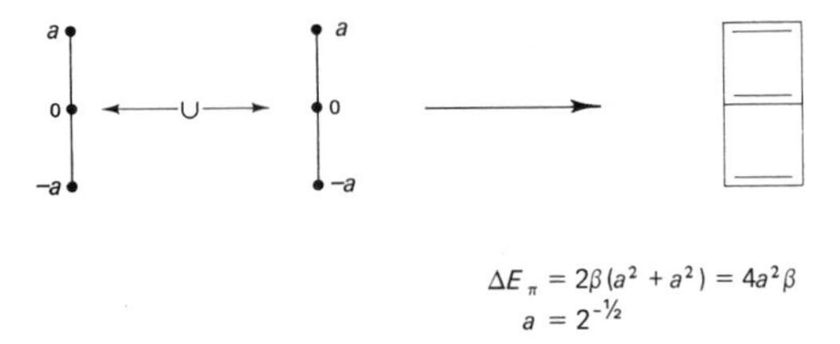

$$\Delta E_\pi = 2\beta(a^2 + a^2) = 4a^2\beta$$
$$a = 2^{-1/2}$$
$$\Delta E_\pi = 2\beta$$

In the case of benzene, $\Delta E_\pi = 4a\beta$, where $a = 1/\sqrt{3}$, so $\Delta E_\pi = (4/\sqrt{3})\beta = 2.31\beta$. Thus, the transition state for process (1) should be favored, as has been experimentally observed by the above authors.

ANSWERS TO PROBLEMS IN CHAPTER III

PROBLEM III.1

(a) *Reference*: C. F. Huebner, E. Donoghue, L. Dorfman, F. A. Stuber, N. Danieli, and E. Wenkert, *Tetrahedron Lett.* p. 1185 (1966).

The reaction can occur in two steps: the first, an ene reaction involving one of the double bonds and an allylic proton in 1,4-cyclohexadiene, is a thermally allowed $[{}_{\sigma}2_s + {}_{\pi}2_s + {}_{\pi}2_s]$ cycloaddition reaction.

COOMe
C
C
H H COOMe
Δ
MeOOC
COOMe
H
(A)

Compound A can then undergo an intramolecular Diels-Alder reaction (thermally allowed) to afford the observed product.

MeOOC COOMe
H
(A)
Δ
H
MeOOC COOMe
≡
COOMe
H
COOMe

(b)

O O O
Δ
$({}_{\pi}2_s + {}_{\pi}2_s + {}_{\pi}2_s)$
O O O

(c) On the basis of the foregoing, we might anticipate the formation of a product analogous to that formed in the reaction of 1,4-cyclohexadiene with dimethyl acetylenedicarboxylate [part (a)].

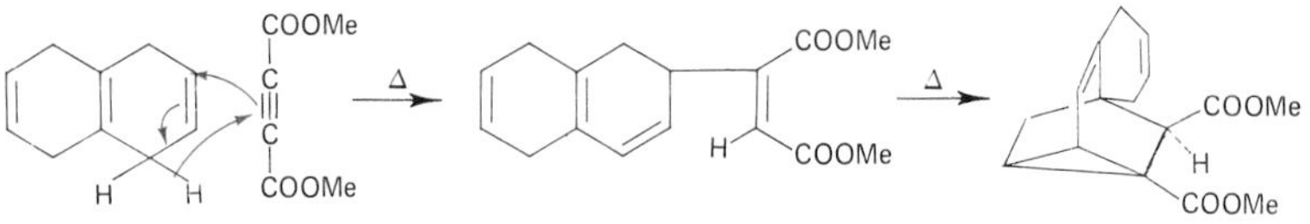

Additionally, two $[_{\pi}2_s + _{\pi}2_s + _{\pi}2_s]$ cycloaddition-type products can be envisioned.

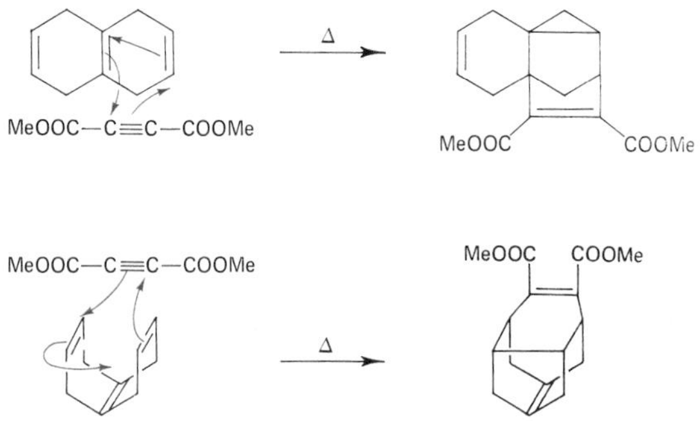

(Other products derived via processes analogous to those depicted above are possible as well.)

PROBLEM III.2

References: R. Askani, *Chem. Ber.* **98**, 3618 (1965); M. Takahashi, Y. Kitahara, I. Murata, T. Nitta, and M. C. Wood, *Tetrahedron Lett.* p. 3387 (1968).

(a) This is a $[_{\pi}2_s + _{\pi}2_s + _{\pi}2_s + _{\pi}2_s]$ cycloaddition reaction (photochemically allowed).

(b)

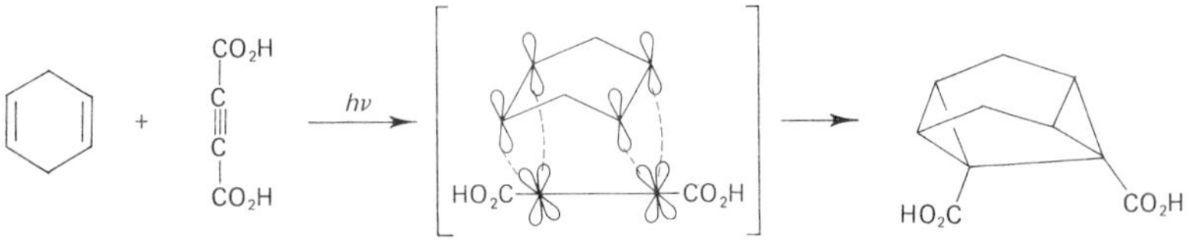

Transition state

(I)

PROBLEM III.3

(a) *Reference*: W. von E. Doering, *U.S. Dept. Com., Office Tech. Serv., PB Rept.* **34**, No. 3, 8 pp., 1960; *Chem. Abstr.* **56**, 5883e (1962). W. von E. Doering and D. W. Wiley, *Tetrahedron* **11**, 183 (1960).

Two thermally allowed processes can be considered: a $[_{\pi}4_s + _{\pi}2_s]$ cycloaddition and a $[_{\pi}8_s + _{\pi}2_s]$ cycloaddition.

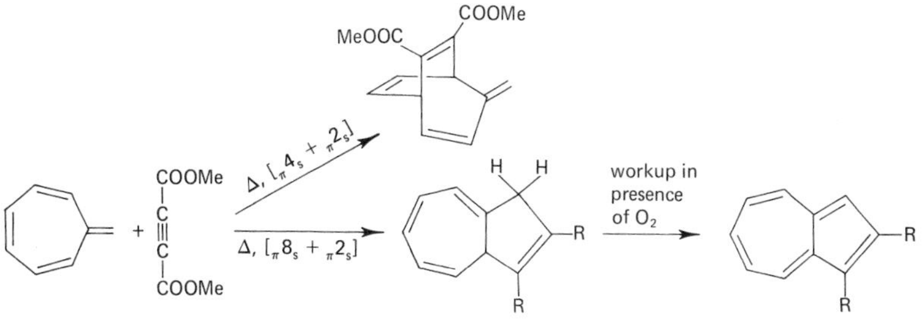

The $[_{\pi}8_s + _{\pi}2_s]$ cycloaddition is actually observed; the observed product, formed by air oxidation of the adduct, is a substituted azulene which possesses some aromatic character.

(b) *Reference*: W. von E. Doering, as cited in Problem III.3 (a).

Two thermally allowed processes merit consideration. The first, which would result in the formation of a 1:1 adduct via a *trans*-$[_{\pi}10_a + _{\pi}2_s]$ cycloaddition, is not observed.

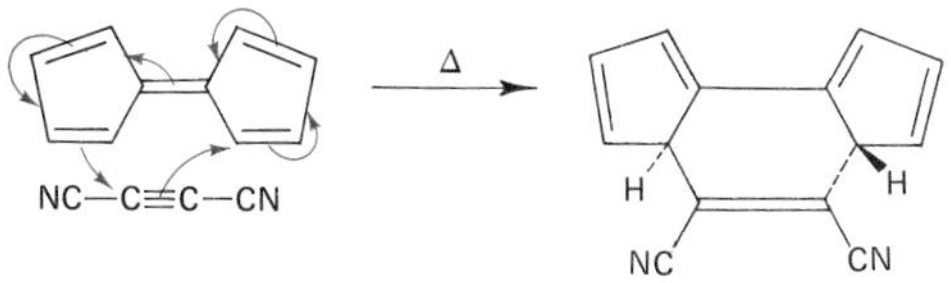

The second process is a double Diels-Alder reaction resulting in the formation of a 1:2 adduct; this is the observed product of the reaction.

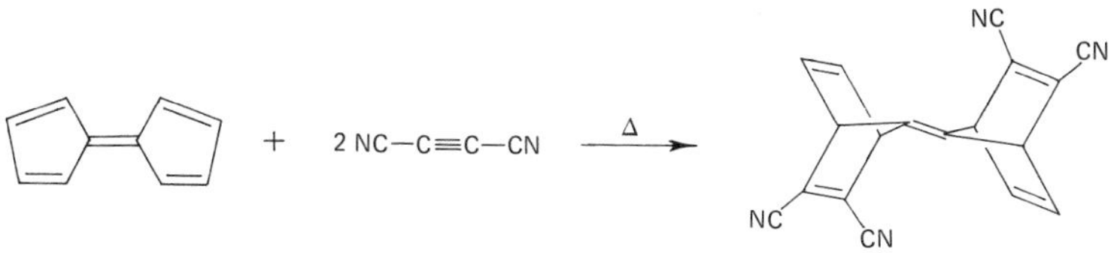

(c) *Reference*: W. von E. Doering *et al.*, cited by R. B. Woodward, *Chem. Soc. Spec. Publ.* No. 21, 217 (1967).

The observed 1:1 adduct is formed via *trans*-$[_{\pi}14_a + {}_{\pi}2_s]$ cycloaddition (thermally allowed) in the manner depicted below:

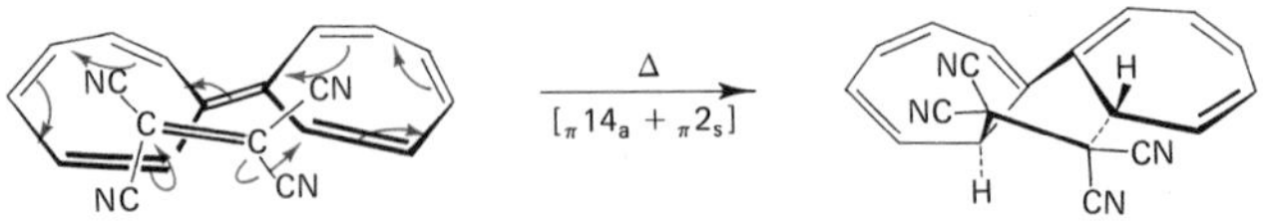

Note that this result contrasts with that obtained with the more rigid pentafulvalene system [part (b)], where strain energy considerations appear to preclude *trans*-$[_{\pi}10_a + {}_{\pi}2_s]$ cycloaddition. No products resulting from an initial $[_{\pi}4_s + {}_{\pi}2_s]$ addition of tetracyanoethylene to the substrate are observed.

(d) *Reference*: J. K. Williams and R. E. Benson, *J. Amer. Chem. Soc.* **84**, 1257 (1962).

Two $[_{\pi}2_s + {}_{\pi}2_s + {}_{\pi}2_s]$ cycloaddition processes can be formally considered:

(1)

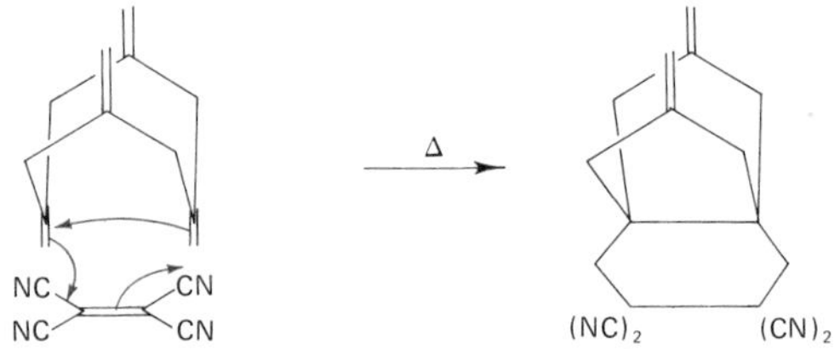

(2)

Consideration of strain effects in the product would argue against the operation of path 2 relative to path 1. The product actually observed is that formed via process 1. Note also that ene reactions are possible in this system. However, no products arising via ene reaction pathways were observed.

(e) *Reference*: A. G. Anastassiou and R. P. Cellura, *J. Chem. Soc.* (D), p. 484 (1970).

The 1:1 adduct can be formed via either of two concerted (thermally allowed) pathways.

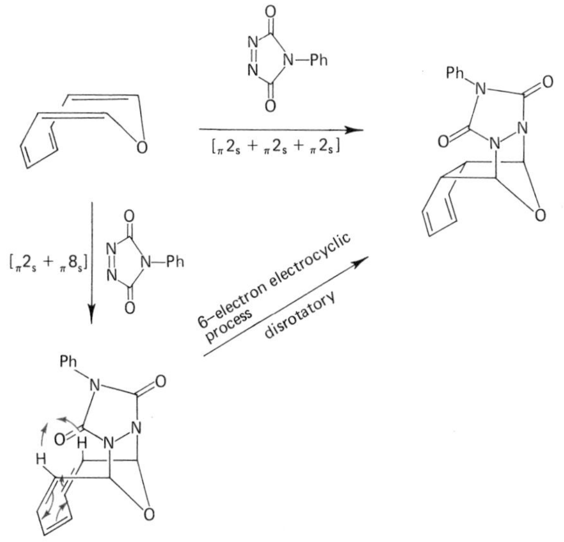

$[_{\pi}2_s + _{\pi}2_s + _{\pi}2_s]$

N—Ph

Ph

O

N

$[_{\pi}2_s + _{\pi}8_s]$

6-electron electrocyclic process disrotatory

(f) *Reference*: O. Sciacovelli, W. von Philipsborn, C. Amith, and D. Ginsburg, *Tetrahedron* **26**, 4589 (1970).

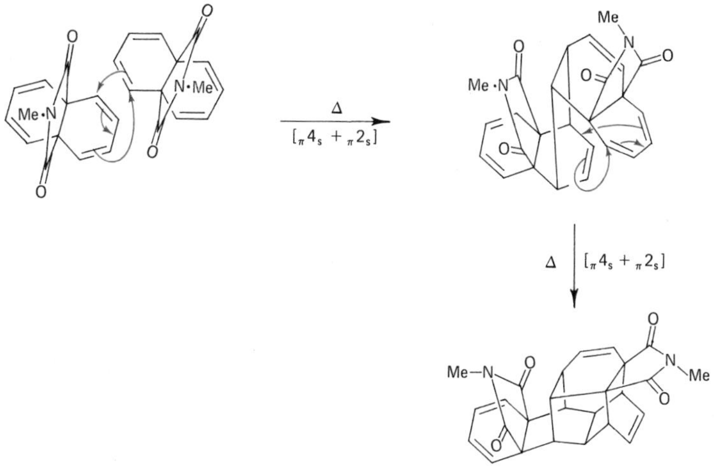

Me·N N·Me

Δ $[_{\pi}4_s + _{\pi}2_s]$

Δ $[_{\pi}4_s + _{\pi}2_s]$

Me—N N—Me

PROBLEM III.4

References: W. H. Okamura and T. W. Osborn, *J. Amer. Chem. Soc.* **92**, 1061 (1970); C. S. Baxter and P. J. Garratt, *ibid.*, p. 1062.

6-electron process disrotatory

Diels-Alder via *endo* transition state

(A)

Note (a)

Diels-Alder

(B)

Note (a): The indicated [1,3] sigmatropic shift is formally a vinylcyclopropane ⟶ cyclopentene rearrangement. This type of rearrangement normally occurs with a high activation energy; whether it proceeds via the symmetry-allowed concerted pathway or via an alternative stepwise process is the subject of much debate (see Problem IV.6 and discussion in the answer to that problem). Note also that the thermally allowed $[_{\pi}4_s + _{\pi}2_s]$ cycloaddition of maleic anhydride to I is not observed.

Δ

PROBLEM III.5

References: D. Bryce-Smith, *J. Chem. Soc.* (D), p. 806 (1969); Professor J. A. Berson, personal communication.

(1) $[_{\sigma}2_s + _{\sigma}2_s + _{\pi}2_s]$ cycloreversion
(2) $[_{\sigma}2_s + _{\sigma}2_s + _{\pi}2_s + _{\pi}2_s]$ cycloreversion
(3) $[_{\sigma}2_s + _{\sigma}2_s]$ cycloreversion
(4) $[_{\sigma}2_s + _{\sigma}2_s]$ cycloreversion

Of the above four processes, only process (1) is symmetry allowed to proceed thermally via a concerted pathway. Symmetry control would therefore facilitate process (1) relative to the other three processes.

PROBLEM III.6

(a) *Reference*: H. E. Zimmerman and G. L. Grunewald, *J. Amer. Chem. Soc.* **86**, 1434 (1964).

R—C≡C—R
(R=CN or COOMe)

Δ
$[_\pi 2_s + _\pi 2_s + _\pi 2_s]$

R R

Δ
$[_\pi 2_s + _\sigma 2_a + _\sigma 2_a]$

s a a
R R
Transition state

H H R R ≡ H R R H

(b) *Reference*: C. G. Krespan, B. C. McKusick, and T. L. Cairns, *J. Amer. Chem. Soc.* **83**, 3428 (1961).

CF_3 C C CF_3

Δ
$[_\pi 4_s + _\pi 2_s]$

CF_3 F_3C
F_3C—C≡C—CF_3

Δ
$[_\pi 2_s + _\pi 2_s + _\pi 2_s]$

CF_3 F_3C F_3C CF_3
(A)

CF_3 F_3C F_3C CF_3

Δ, symmetry "forbidden"
$[_\sigma 2_s + _\pi 2_s + _\sigma 2_s + _\pi 2_s + _\sigma 2_a]$

F_3C CF_3 s s s a F_3C s CF_3
Transition state

CF_3 F_3C H H F_3C CF_3
(B)

(B) ≡ F_3C CF_3 F_3C CF_3

Note that compound A cannot rearrange thermally to B directly via a symmetry-allowed concerted pathway; the formation of A itself from reactants can, however, occur in two symmetry-allowed steps.

The observed products, I and bis(trifluoromethyl)ethylene, can be obtained thermally from B via a $[_\sigma 2_s + {}_\sigma 2_s + {}_\pi 2_s]$ cycloaddition (oxidation–reduction) reaction.

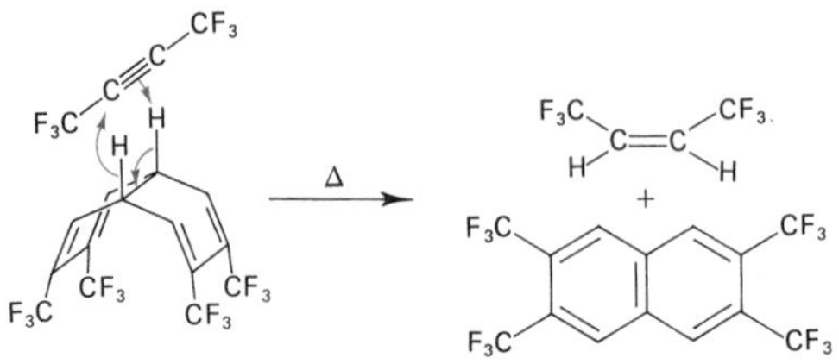

Although Cairns did not report the stereochemistry of the bis(trifluoromethyl)-ethylene, we would expect it to possess the *cis* configuration on the basis of the above mechanism.

Additionally, the product naphthalene has been found capable of further reaction with excess bis(trifluoromethyl)acetylene.

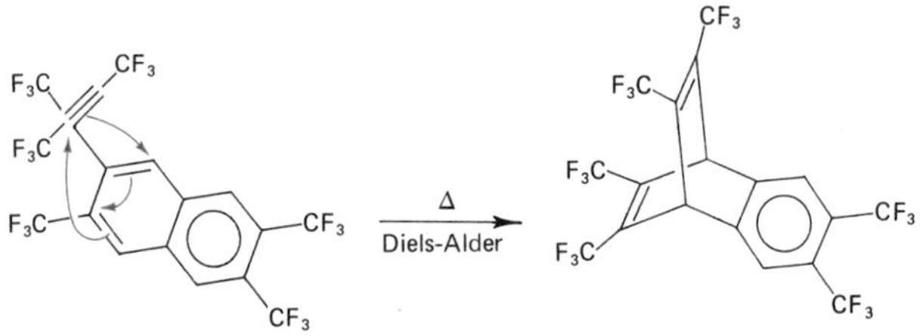

One additional comment is in order: Zimmerman's observation [part (a) of this question] that an intermediate analogous to A affords a 1,2-disubstituted naphthalene as the final product suggests the improbability of the intermediacy of A in the reaction studied by Cairns. The reader should consult the paper by Cairns and co-workers to obtain suggestions for alternative (stepwise) mechanisms in this reaction.

PROBLEM III.7

Reference: K. Takatsuki, I. Murata, and Y. Kitahara, *Bull. Chem. Soc. Jap.* **43**, 966 (1970).

The thermally produced dimer is formed in two steps: a symmetry-allowed $[_\pi 6_s + {}_\pi 4_s]$ cycloaddition followed by an intramolecular Diels-Alder reaction.

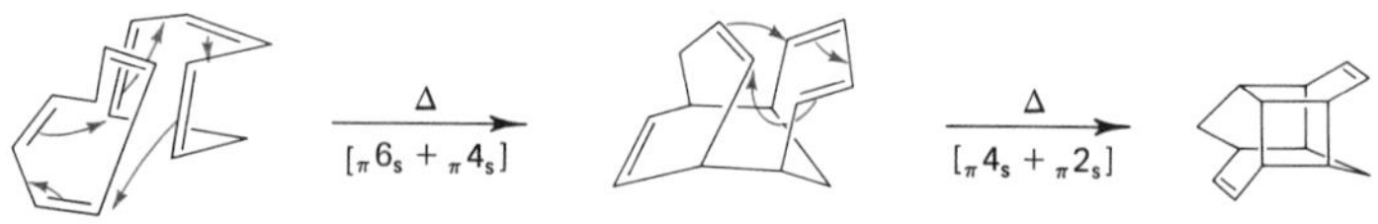

PROBLEM III.8

Reference: P. M. Weintraub. *J. Chem. Soc.* (D), p. 760 (1970).

The first step is a 1,3-dipolar cycloaddition reaction. These cycloaddition reactions have been studed in detail by Professor Rolf Huisgen, Institute for Organic Chemistry, University of Munich, Germany.* These 1,3-dipolar cycloadditions are isoelectronic with $[_{\pi}4_s + _{\pi}2_s]$ cycloadditions and are therefore symmetry-allowed thermal reactions. This step is followed by a 1,3-dipolar cycloreversion with consequent elimination of CO_2. A final 1,3-dipolar cycloaddition reaction produces the observed product, II.

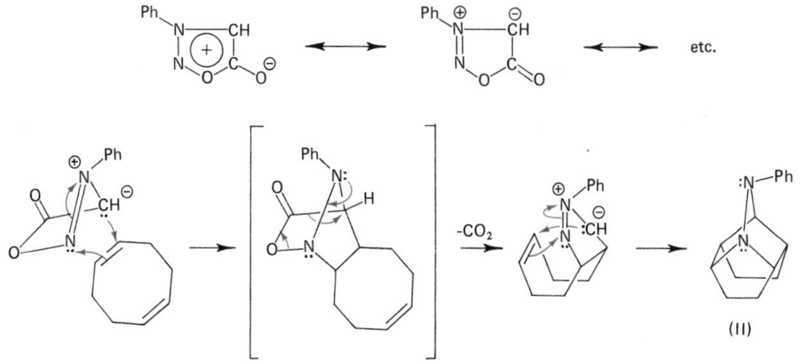

etc.

$-CO_2$

(II)

PROBLEM III.9

Reference: W. L. Mock, *J. Amer. Chem. Soc.* **92**, 3807 (1970).

Process (1), a six-electron thermal cycloreversion, is a symmetry-allowed linear cheletropic process. Process (2), an eight-electron thermal cycloreversion, is predicted on the basis of orbital symmetry considerations to occur as a *nonlinear* cheletropic process.† We can analyze these reactions through the reverse of the

(1) $[_{\pi}4_s + _{\omega}2_s]$ Δ Δ SO_2 (I)

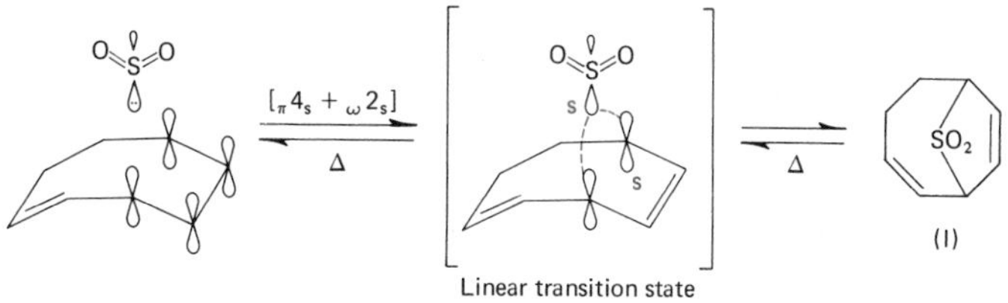

Linear transition state

*For a review, see R. Huisgen, *Angew. Chem. Int. Ed. Engl.* **2**, 565, 633 (1963).

†For a discussion of linear vs. nonlinear cheletropic processes, see R. B. Woodward and R. W. Hoffmann, "The Conservation of Orbital Symmetry," pp. 152–163. Academic Press, New York, 1970. For a definition of cheletropic process, see footnote on page 160.

(2)

$[_\pi 6_s + _\omega 2_a]$

Δ

Nonlinear transition state

processes shown (i.e., as cycloadditions of SO_2 to 1,3,5-cyclooctatriene). The transition states of retrograde reactions (1) and (2) are seen to result from linear $[_\pi 4_s + _\omega 2_s]$ and nonlinear $[_\pi 6_s + _\omega 2_a]$ cheletropic processes, respectively.

Mock presents evidence and arguments which suggest that the observed $\Delta\Delta F^*$ reflects the orbital symmetry-imposed preference for the linear [path (1)] over the nonlinear [path (2)] cheletropic process. However the observation that the $\Delta\Delta F^*$ value arises primarily through the difference in the *entropy* terms for reactions (1) and (2) suggests that serious consideration should be given to the possibility of a gross change in mechanism in proceeding from reaction (1) to reaction (2). Conclusions based on direct comparison of reactions (1) and (2) as "concerted" processes are, therefore, tenuous.

PROBLEM III.10

Reference: L. A. Paquette and L. M. Leichter, *J. Amer. Chem. Soc.* **92**, 1765 (1970).

$Fe(CO)_3$ $\xrightarrow{Ce^{3+}}$ [ ] $\xrightarrow{\Delta}$ (A) $\xrightarrow{h\nu,\ -N_2}$ (B) $\xrightarrow{100°C}$ (C); (A) $\xrightarrow{130°C}$ (C)

Compound A is formed by a Diels-Alder reaction in which the cyclobutadiene moiety fulfills the role of dienophile (note the *endo* configuration of the adduct). The photoconversion A ⟶ B is a symmetry-allowed $[_\sigma 2_s + _\sigma 2_s]$ cycloreversion.

The thermal process A $\longrightarrow$ C is a symmetry-allowed $[_{\pi}2_s + _{\sigma}2_s + _{\sigma}2_s]$ intramolecular cycloaddition; however, the corresponding process B $\longrightarrow$ C is symmetry "forbidden."

Me Me Ph Ph N N (A) Δ $[_{\pi}2_s + _{\sigma}2_s + _{\sigma}2_s]$ Transition state Me Me Ph Ph (C)

Me Me Ph Ph N N (A) $h\nu, -N_2$ $[_{\sigma}2_s + _{\sigma}2_s]$ Me Me Ph Ph (B) Δ $[_{\sigma}2_s + _{\pi}2_s]$ symmetry "forbidden" (C)

PROBLEM III.11

Reference: K. N. Houk and R. B. Woodward, *J. Amer. Chem. Soc.* **92**, 4143 (1970).

Adduct A is formed via an ordinary Diels-Alder reaction (*endo* transition state preferred) in which cycloheptatriene functions as the dienophile.

Ph Me O Ph Me Δ $[_{\pi}4_s + _{\pi}2_s]$ O Me Ph Ph Me (A)

Adduct B is formed via a $[_{\pi}6_s + _{\pi}4_s]$ process (thermally allowed, but proceeding via the *exo* transition state).

Ph Me O Ph Me Δ $[_{\pi}6_s + _{\pi}4_s]$ O Me Ph Ph Me (B)

Adduct C is formed from A via a [3,3] sigmatropic shift (Cope rearrangement).

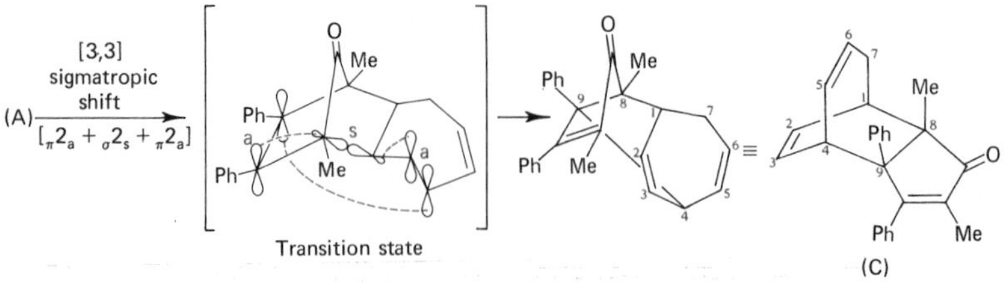

(A) [3,3] sigmatropic shift $[_{\pi}2_a + _{\sigma}2_s + _{\pi}2_a]$ Transition state (C)

Adduct D can be formed from B via a [1,3] sigmatropic shift (symmetry "forbidden" to occur thermally in a concerted fashion), followed by an intramolecular Diels-Alder reaction.

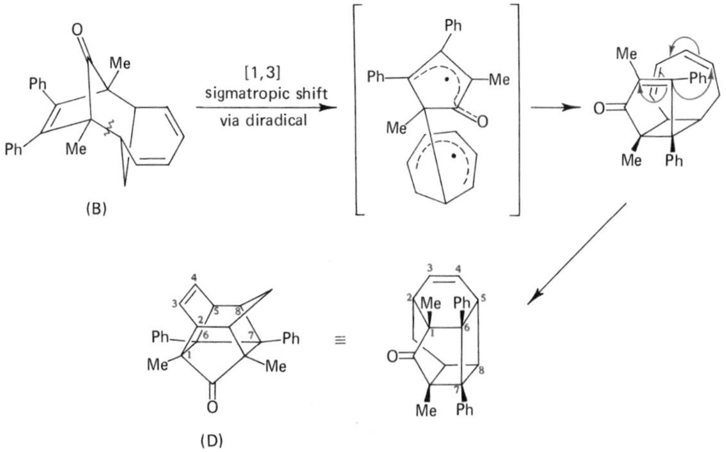

(B) [1,3] sigmatropic shift via diradical (D)

Compound E arises from thermal decarbonylation of A via a $[_{\sigma}2_s + _{\sigma}2_s + _{\pi}2_s]$ cycloreversion, followed by a sequence of [1,5] hydrogen shifts.

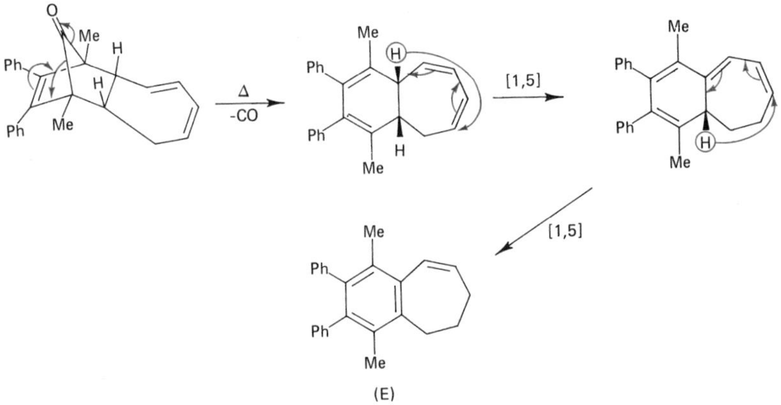

Δ -CO [1,5] [1,5] (E)

Compound F is simply a 2:1 Diels-Alder adduct of diene I with the dienophile, cycloheptatriene.

PROBLEM III.12

Reference: K. N. Houk, *Tetrahedron Lett*. p. 2621 (1970).

(a) The Diels-Alder reaction with the dienophile cyclopentadiene proceeds preferentially via the *endo* transition state in accord with the Alder rule of "maximum accumulation of unsaturation." The reaction with cyclopentene as dienophile lacks the additional unsaturation that was present in the reaction with cyclopentadiene. There are no longer "secondary attractive forces"* to favor *endo* addition over the *exo* mode, and the reaction therefore loses its stereoselectivity.

(b) Compounds A and B can interconvert via a [3,3] sigmatropic rearrangement (Cope rearrangement).

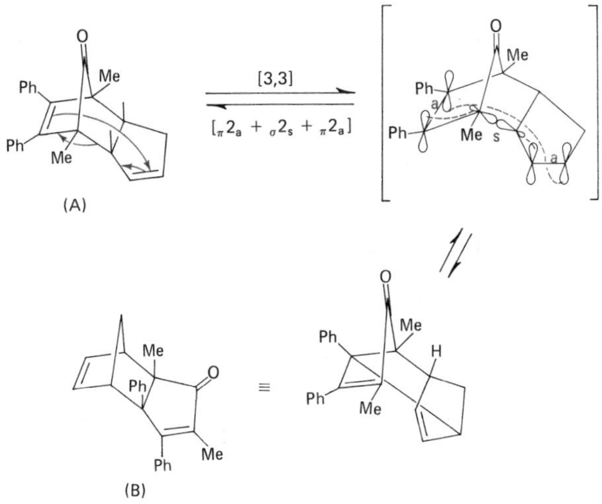

(A) $\underset{[_{\pi}2_a + {}_{\sigma}2_s + {}_{\pi}2_a]}{\overset{[3,3]}{\rightleftharpoons}}$

(B) ≡

PROBLEM III.13

Reference: W. T. Ford, R. Radue, and J. A. Walker, *J. Chem. Soc.* (D), p. 966 (1970).

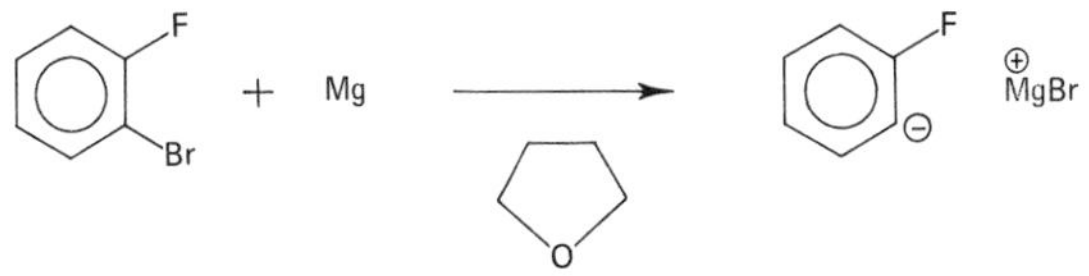

* For a discussion, see text material relating to Fig. II.15. See also Problem II.14.

F ⊕ MgBr + → F H + ⊖ ⊕ MgBr (B)

F → + $F^{\ominus}$

(B) ⊕ MgBr + $[_{\pi}4_s + {}_{\pi}2_s]$ H ⊖ ⊕ MgBr

D_2O H_2O

H D (A′) H H (A)

The fact that deuterium uptake is observed upon hydrolysis in D_2O (to form A′) confirms that A is formed by a $[_{\pi}4_s + {}_{\pi}2_s]$ cycloaddition of benzyne to the *anion* B (or to C_9H_7MgBr) rather than to isoindene itself.

ANSWERS TO PROBLEMS IN CHAPTER IV

PROBLEM IV.1

(a)

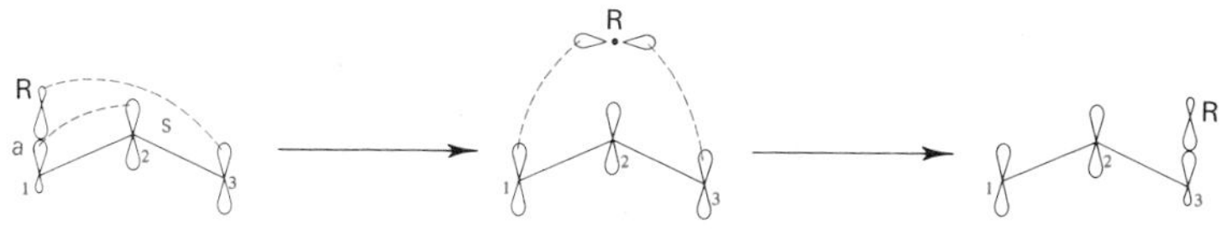

Transition state

$[_{\pi}2_s + _{\sigma}2_a]$

Thermally allowed

The new bond to R is formed from the backside in the example shown above. Clearly, R suffers *inversion* of configuration in undergoing the rearrangement. A similar analysis of the corresponding photochemical process reveals that the rearrangement should now occur with *retention* of configuration of the migrating group, R; the reader should verify this.

(b) The reader can verify that the results are now just the opposite of those obtained above from our analysis of the [1,3] sigmatropic rearrangement: i.e., the thermal suprafacial process occurs with retention and the corresponding photochemical process occurs with inversion of configuration at R.

Of course, any of the other theoretical treatments when applied to this problem should afford the same predictions. Thus, for the [1,5] sigmatropic rearrangement, the basis sets for migration with retention and with inversion, respectively, are shown below (red = positive lobe, black = negative lobe):

Transition state, migration with retention

Transition state, migration with inversion

The transition state for migration with retention is seen to be a Hückel system and is isoconjugate with benzene, a Hückel-aromatic species. The reaction is thus thermally allowed. There is one sign inversion in the transition state for migration with inversion; this is a Möbius system. The transition state is again isoconjugate with benzene but is antiaromatic. The reaction is thus photochemically allowed and thermally forbidden.

PROBLEM IV.2

(a) *References*: J. A. Berson and G. L. Nelson, *J. Amer. Chem. Soc.* **89**, 5503 (1967); J. A. Berson, *Accounts Chem. Res.* **1**, 152 (1968); J. A. Berson and R. S. Wood, *J. Amer. Chem. Soc.* **89**, 1043 (1967); R. C. Cookson, *Quart. Revs.* **22**, 423 (1968). For another view of the mechanism of this rearrangement, see H. E. Zimmerman, D. S. Crumrine, D. Döpp, and P. S. Huyffer, *J. Amer. Chem. Soc.* **91**, 434 (1969).

This is a [1,3] suprafacial sigmatropic rearrangement. As we have seen in Problem IV.1, this reaction is predicted to proceed thermally with inversion of configuration of the migrating group. (Note that D and OAc are *trans* in the reactant and *cis* in the product.) A reasonable mechanism is shown below:

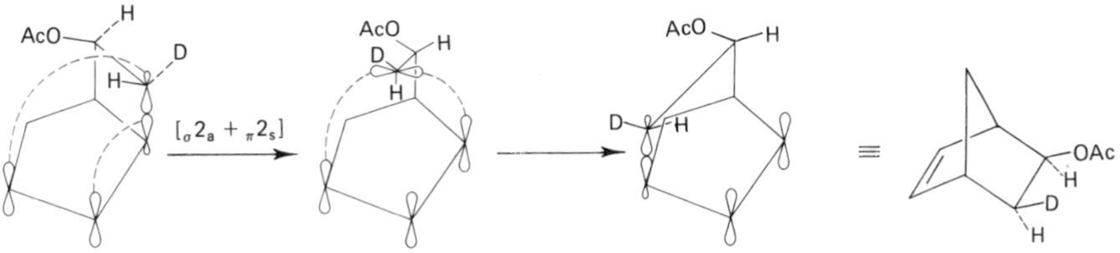

(I)

(b) *Reference*: J. A. Berson and G. L. Nelson, *J. Amer. Chem. Soc.* **92**, 1096 (1970).

The results suggest that compound I rearranges predominantly by a concerted mechanism analogous to that suggested in part (a) of this question. However, compound II rearranges via a *stepwise* mechanism. This change in mechanism is suggested both by the difference between the stereochemistry of the predominant migration process for I and II, and by the *exo*, *endo* epimerization data. An examination of the symmetry-allowed processes reveals that severe nonbonded interactions develop in the transition state for rearrangement of II, since the methyl group attached to the migrating carbon atom would be forced into the face of the five-membered ring (see intermediate I in the answer to Problem IV.2; in the rearrangement of II, methyl rather than hydrogen will be thrust into the ring). The symmetry-allowed pathway for rearrangement of II is thereby disfavored, and the stepwise process occurs.

PROBLEM IV.3

References: J. A. Berson, *Accounts Chem. Res.* **1**, 152 (1968); J. A. Berson and M. R. Willcott, III, *Rec. Chem. Progr.* **27**, 139 (1966); R. B. Woodward and R. Hoffmann, "The Conservation of Orbital Symmetry," pp. 124–125. Academic Press, New York, 1970.

When the sequence of [1,5] sigmatropic rearrangements and six-electron electrocyclic processes is carried through for the thermally allowed case (retention at C_7), it is seen that the optical purity of the starting material is preserved throughout the rearrangement (the reader should verify this). However, operation of the "forbidden" mechanism results in racemization. This is readily seen by examination of the following circuit:

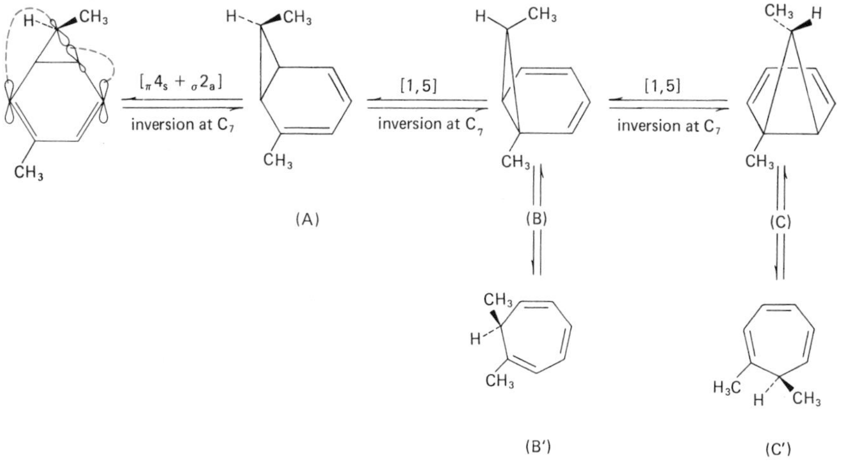

$[{}_{\pi}4_s + {}_{\sigma}2_a]$ inversion at C_7 (A); [1,5] inversion at C_7 (B) ⇌ (B′); [1,5] inversion at C_7 (C) ⇌ (C′)

It can be seen that B and C are enantiomers, as are their ring-opened counterparts, B′ and C′. Hence, the operation of the thermally "forbidden" mechanism would result in *destruction* of optical activity (racemization) in contrast to the results obtained from the thermally allowed process. The reader can verify that this same result obtains upon either clockwise or counterclockwise revolution of the norcaradiene interconversion circuit; neither mechanism distinguishes clockwise from counterclockwise revolution.

PROBLEM IV.4

The slither mechanism is stereochemically equivalent to the [1,5] suprafacial sigmatropic shift with inversion. Thus, racemization would occur if that mechanism were followed. The reader may confirm that the intermediates B and C (answer to Problem IV.3), which are enantiomers, are easily interconverted by the slither mechanism.

PROBLEM IV.5

References: T. M. Brennan and R. K. Hill, *J. Amer. Chem. Soc.* **90**, 5614 (1968); H. E. Zimmerman and D. S. Crumrine, *ibid.* **90**, 5612 (1968); H. E. Zimmerman, D. S. Crumrine, D. Döpp, and P. S. Huyffer, *ibid.* **91**, 434 (1969).

(a) (1) slither

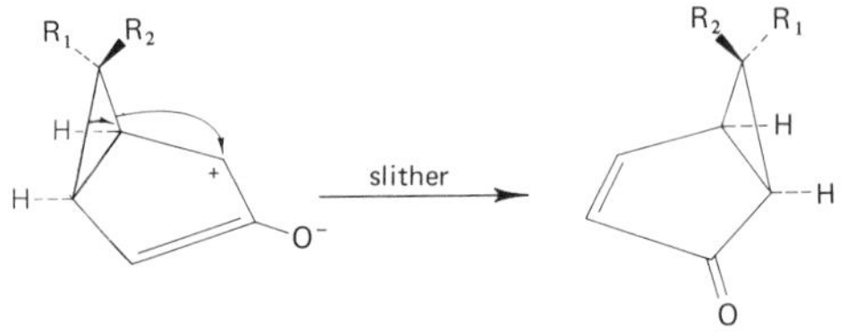

(2) [1,4] shift with retention

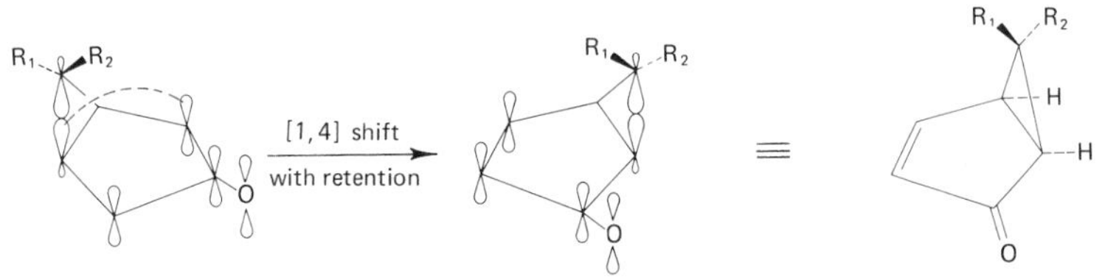

(3) [1,4] shift with inversion

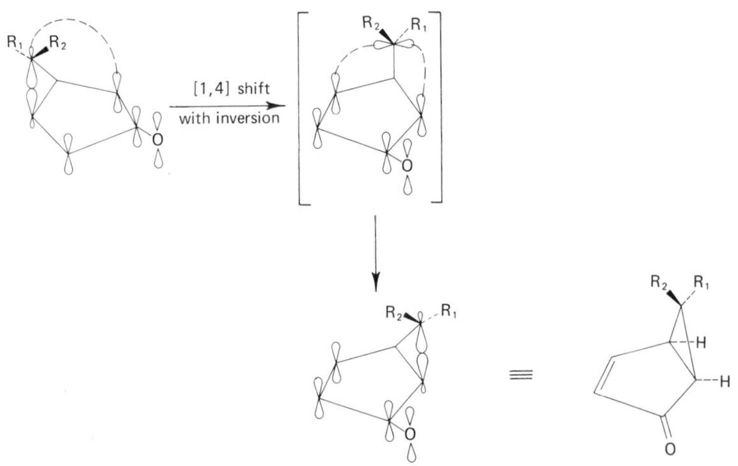

(b) If one makes the (rather drastic) assumption that one can neglect the cross-conjugated O^-, this is a suprafacial [1,4] shift in a carbonium ion; to be allowed, it should be $[_\sigma 2_a + {}_\pi 2_s]$—the migration should proceed with inversion. In PMO terms, the transition state is isoconjugate with cyclopentenyl cation (if one neglects the O^-), and it is aromatic only if Möbius. That condition requires inversion at the migrating group, also. If one includes the oxygen in the PMO treatment, the transition state is isoconjugate with cyclopentadienone, which would be aromatic only if Möbius, so inversion of the migrating group is again predicted.

(c) Both results are consistent with a suprafacial [1,4] sigmatropic shift proceeding with inversion, as predicted. The slither mechanism also accommodates the results.

PROBLEM IV.6

(a) $[_\sigma 2_a + {}_\pi 2_s]$ and $[_\sigma 2_s + {}_\pi 2_a]$

(b) (1) $[_\sigma 2_a + {}_\pi 2_s]$

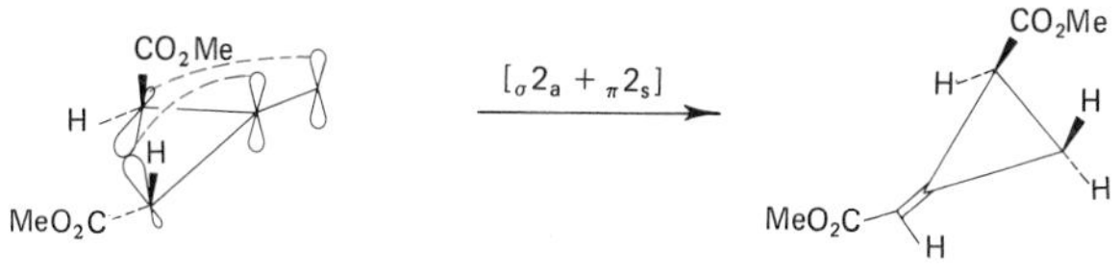

(2) $[_\sigma 2_a + {}_\pi 2_s]$

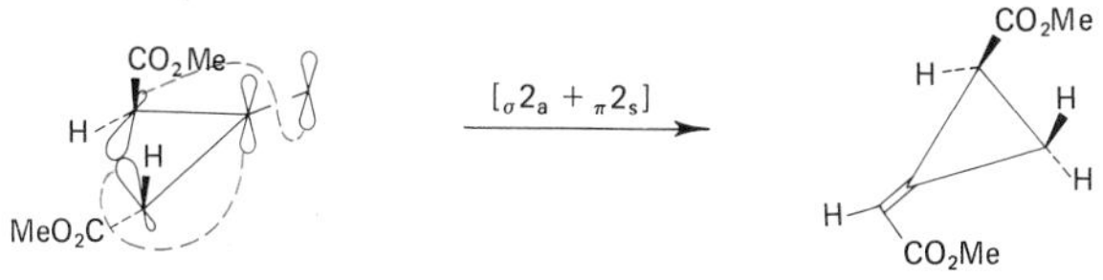

(3) $[_\sigma 2_a + {}_\pi 2_s]$

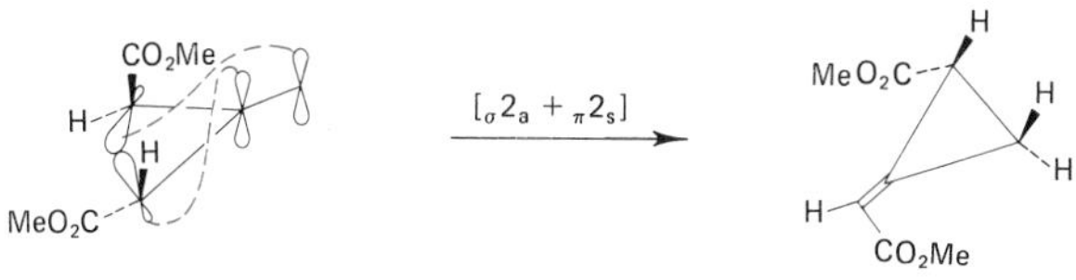

(4) $[_\sigma 2_a + {}_\pi 2_s]$

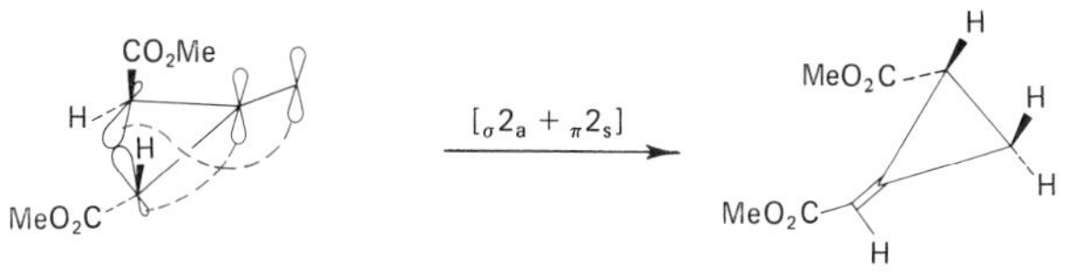

(c) Processes (1) and (2) should be (and actually are) favored. Pathways (3) and (4), though formally allowed, require a 90° rotation of the orbital to which the new cyclopropane bond will be formed before bonding of it with the methylene *p* orbital can occur, due to the intervening framework of sigma-bonds. In the case cited here, a nonconcerted mechanism could also produce the observed results. For a discussion of this problem and a suggested method of determining whether the methylenecyclopropane rearrangement (of which Feist's ester is one example), is concerted, see W. von E. Doering and H. D. Roth, *Tetrahedron* **26**, 2825 (1970).

PROBLEM IV.7

Reference: T. Miyashi, M. Nitta, and T. Mukai, *Tetrahedron Lett.* p. 3433 (1967). In the rearrangement, the bond is broken at C_4—C_7 and reformed at C_2—C_5.

(a)

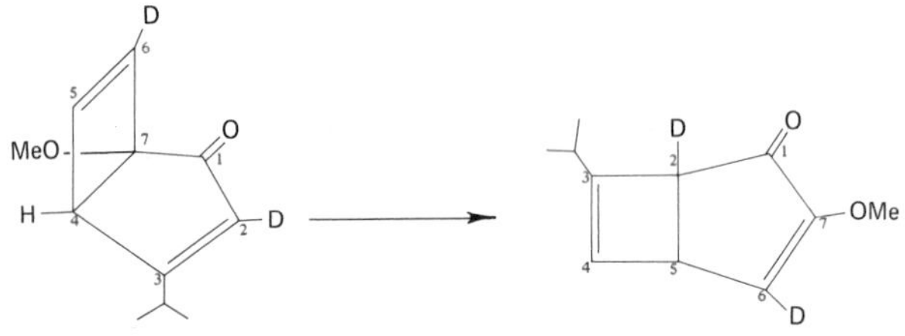

In orbital terms:

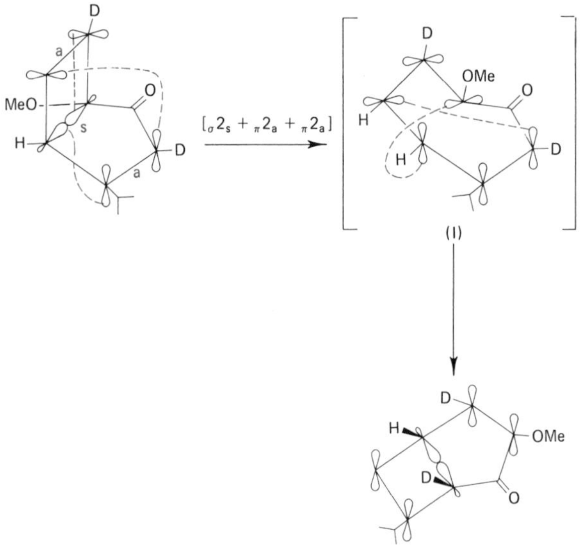

$[_{\sigma}2_s + _{\pi}2_a + _{\pi}2_a]$

(I)

(b) The process is *antara–antara* with respect to the pi-system (see "transition state" I).

(c) The process is particularly worthy of note because it is *antara–antara* with respect to the pi-systems. This is extremely rare, and requires special geometric disposition of the orbitals involved, a criterion met in the above example.

A more recent study by Baldwin and Kaplan* has cast doubt upon the necessity of postulating a direct [3,3] *antara–antara* sigmatropic rearrangement for the process depicted in part (a). They first demonstrated that the analogous rearrangement could occur in an unsubstituted carbocyclic system of suitable geometry.

(1) Δ, 180°C

However, the corresponding process was *not* observed in the following system, although this system meets the geometrical requirements, making it potentially capable of direct [3,3] *antara–antara* sigmatropic rearrangement.

(2) Δ (not observed, even at temperatures as high as 450°C)

Their failure to observe the thermal rearrangement depicted by (2) led Baldwin and Kaplan to propose an alternative to the direct [3,3] sigmatropic shift mechanism to account for the observed rearrangement depicted by (1). Their alternative two-step mechanism for process (1) involves thermal conrotatory electrocyclic opening of the cyclobutene ring, resulting in the formation of *cis, trans, cis*-cyclooctatriene; in a second step, electrocyclic ring closure occurs to give the observed rearrangement product.

Δ $[_{\pi}2_s + _{\sigma}2_a]$ conrotatory

Not isolated

Δ $[_{\pi}2_s + _{\pi}2_a]$

*J. E. Baldwin and M. S. Kaplan, *J. Chem. Soc.* (D), p. 1560 (1970).

This same alternative two-step mechanism can explain the rearrangement observed by Mukai and his co-workers [part (a) of this question].

D OMe H O D

Δ $[_\pi 2_s + _\sigma 2_a]$ conrotatory

MeO O D H a s D

Not isolated

Δ $[_\pi 2_s + _\pi 2_a]$

MeO O D H D

PROBLEM IV.8

(a) *Reference*: J. Almy and D. J. Cram, *J. Amer. Chem. Soc.* **92**, 4316 (1970).

CH_3 H_a H_b CMe_3

[1,5]

CH_3 H_a H_b CMe_3 (I)

[1,5], H_a

CH_3 H_b H_a CMe_3 (+)-(II)

[1,5], H_b

CH_3 H_a H_b CMe_3 (−)-(II)

H_a and H_b in the achiral intermediate I are equivalent. Migration of either is equally likely, so this mechanistic route leads to racemic II.

(b)

(1) CH_3 H H CMe_3

[1,3]

CH_3 H Me_3C H (+)-II

(2)

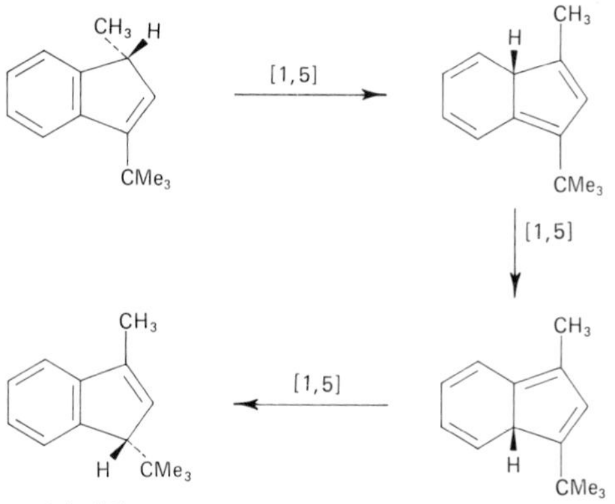

[1,5]

[1,5]

[1,5]

(+)-(II)

Pathway (1) is "forbidden"; (2) is allowed. This latter is therefore preferred.

PROBLEM IV.9

Reference: R. Cargill, B. M. Gimarc, D. M. Pond, T. Y. King, A. B. Sears, and M. R. Willcott, *J. Amer. Chem. Soc.* **92**, 3809 (1970).

(a) If concerted,

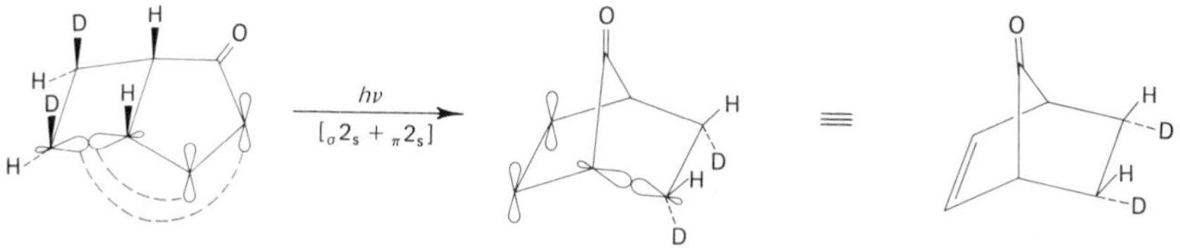

$h\nu$ $[_{\sigma}2_s + _{\pi}2_s]$

Retention of the migrating group. Stereospecific.

(b)

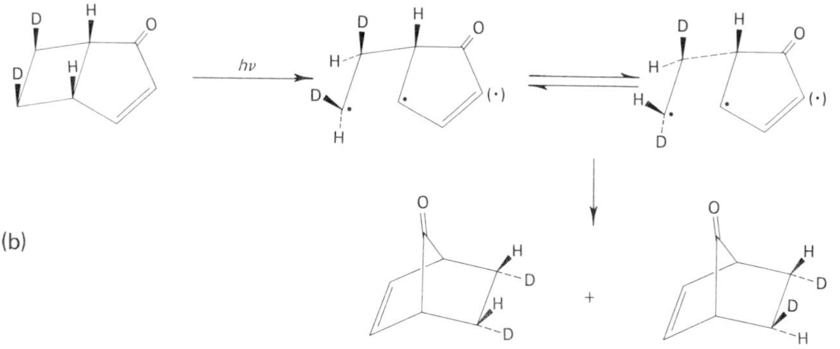

$h\nu$

Quite possibly, the photochemical reaction proceeds through a 1,4-diradical-like intermediate, which allows the migrating atom to equilibrate, as shown.

ANSWERS TO PROBLEMS IN CHAPTER V

PROBLEM V.1

Reference: W. von E. Doering and W. R. Roth, *Angew. Chem. Int. Ed. Engl.* **2**, 115 (1963).

Once deuterium is substituted at a vinyl position in bullvalene, the deuterium can became statistically scrambled throughout the molecule. A complete accounting of this process is presented by Doering and Roth.

PROBLEM V.2

(a) *Reference*: G. Schröder, J. F. M. Oth, and R. Merenyi, *Angew. Chem. Int. Ed. Engl.* **4**, 752 (1965).

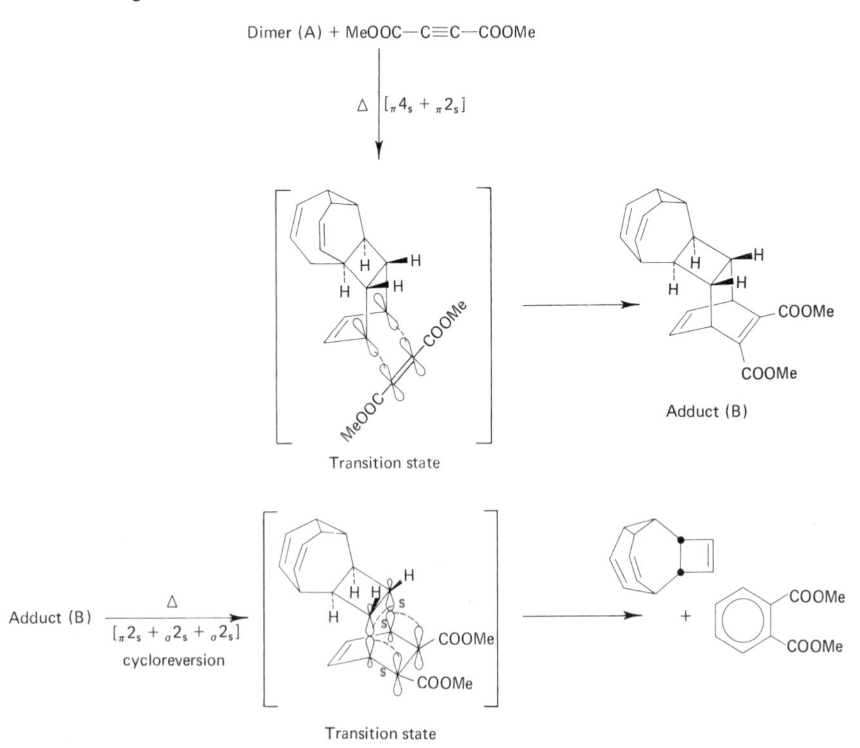

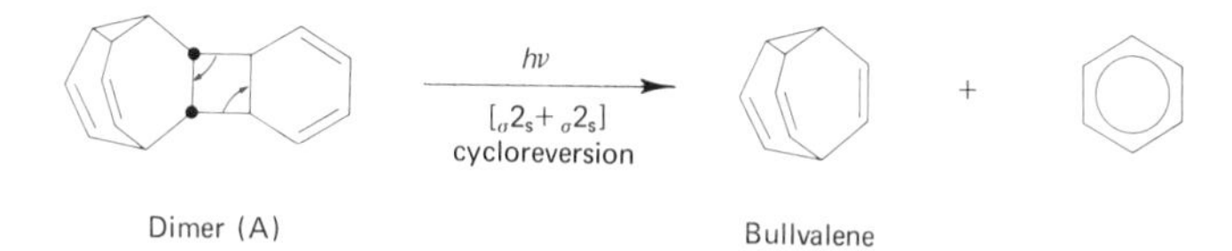

(b) *References*: W. von E. Doering and J. W. Rosenthal, *J. Amer. Chem. Soc.* **88**, 2078 (1966); W. von E. Doering and J. W. Rosenthal, *Tetrahedron Lett.* p. 349 (1967).

The intermediacy of the $[_\pi 4_s + _\pi 4_s]$ intramolecular cycloaddition product (photochemically allowed) has been proposed; the ultimate formation of bullvalene is rationalized as follows:

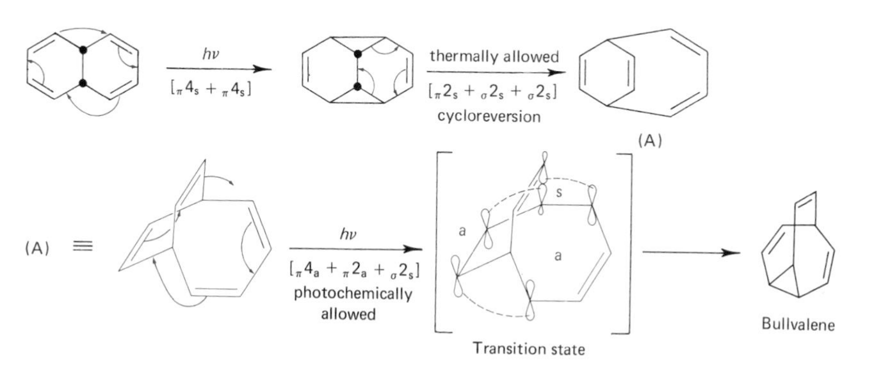

PROBLEM V.3

(a) *Reference*: H. E. Zimmerman and G. L. Grunewald, *J. Amer. Chem. Soc.* **88**, 183 (1966).

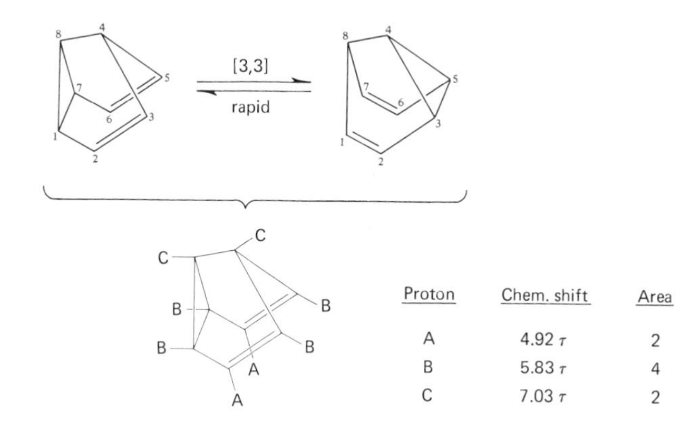

Proton	Chem. shift	Area
A	4.92 τ	2
B	5.83 τ	4
C	7.03 τ	2

(b) *Reference*: H. E. Zimmerman and H. Iwamura, *J. Amer. Chem. Soc.* **92**, 2015 (1970).

A reasonable mechanism for this reaction (if concerted) is outlined below:

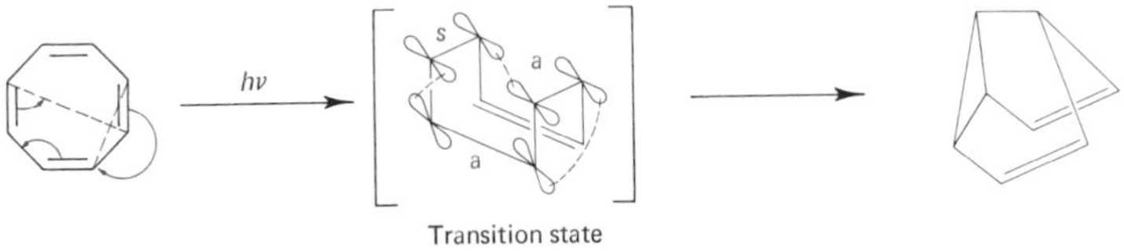

Transition state

This is a $[_{\pi}2_a + {_{\pi}2_a} + {_{\pi}2_s}]$ intramolecular cycloaddition reaction, which is symmetry "forbidden" to occur photochemically in a concerted fashion. The function of the sensitizer is thus to permit direct formation of the triplet (diradical), which can then undergo reaction via a stepwise process. The photochemical process, then, is best pictured as proceeding in a stepwise fashion.

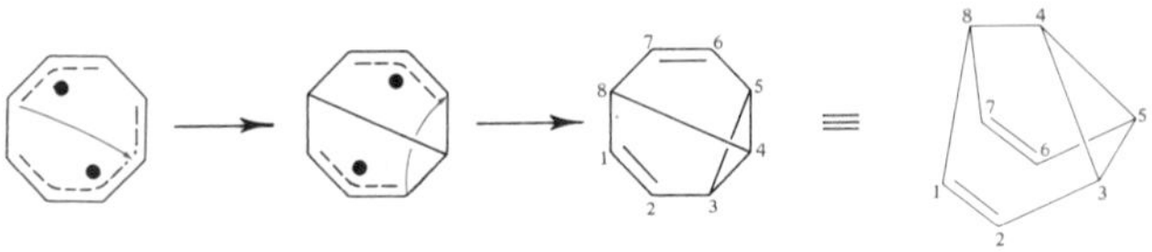

It is worthwhile mentioning in this connection that *thermal* reactions have been studied for reactions closely related to the cyclooctatetraene ⟶ semibullvalene conversion. Octamethylcyclooctatetraene has been converted thermally to octamethylsemibullvalene [reaction (1)*] and dibenzo[*a,c*]cyclooctatetraene has been postulated to readily undergo the conversion shown in reaction (2).†

(1) Δ symmetry allowed

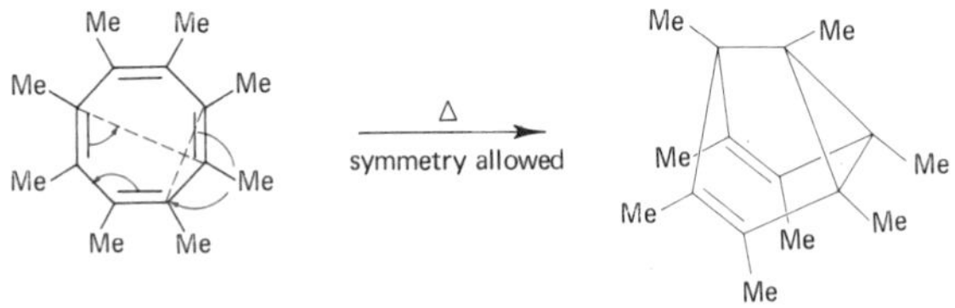

(2) Δ symmetry allowed

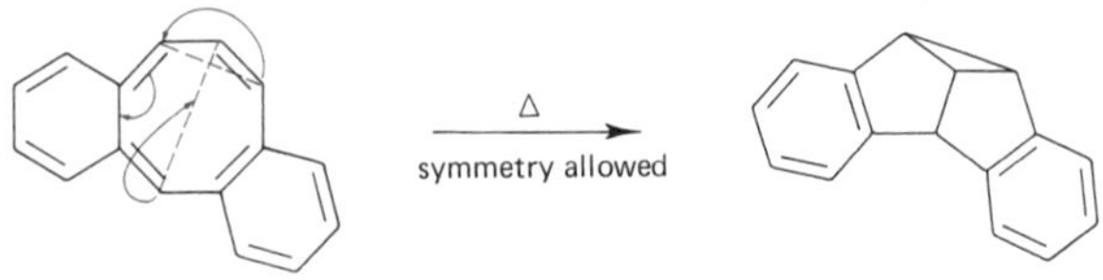

* R. Criegee and R. Askani, *Angew. Chem. Int. Ed. Engl.* **7**, 537 (1968).

† G. F. Emerson, L. Watts, and R. Pettit, *J. Amer. Chem. Soc.* **87**, 131 (1965); W. Merk and R. Pettit, *ibid.* **89**, 4787 (1967).

(c) *References*: H. E. Zimmerman and G. L. Grunewald, *J. Amer. Chem. Soc.* **88**, 183 (1966); H. E. Zimmerman, R. W. Binkley, R. S. Givens, G. L. Grunewald and M. A. Sherwin, *ibid.* **91**, 3316 (1969).

The acetone-sensitized photochemical conversion of barrelene to semibullvalene is a $[_\pi 2_a + {_\pi 2_s} + {_\sigma 2_a}]$ process. It is symmetry forbidden to occur photochemically in a concerted fashion, and it therefore probably occurs via a stepwise process. [See Problem V.3(b); however, see footnote 11 in *J. Amer. Chem. Soc.* **88**, 183 (1966).] The direct photolysis of barrelene to cyclooctatetraene may proceed photochemically via a symmetry-allowed pathway. These results are summarized below:

$h\nu$, symmetry forbidden

Transition state

$h\nu$, $[_\pi 2_s + {_\pi 2_s}]$ — Δ, $[_\pi 2_s + {_\sigma 2_s} + {_\sigma 2_s}]$ — $h\nu$ or Δ

The mechanism shown above for the barrelene ⟶ cyclooctatetraene rearrangement is exactly analogous to the mechanism proposed by Zimmerman and co-workers* to account for their observed photochemical conversion of benzobarrelene to benzocyclooctatetraene. Another symmetry-allowed mechanism can as well be envisioned to account for the photochemical rearrangement of barrelene to cyclooctatetraene.

$h\nu$, $[_\pi 2_s + {_\sigma 2_s}]$

Transition state

$h\nu$, $[_\pi 2_s + {_\sigma 2_s}]$

Transition state

*H. E. Zimmerman, R. S. Givens, and R. M. Pagni, *J. Amer. Chem. Soc.* **90**, 6096 (1968).

Note that this last reaction can also occur in a symmetry-allowed *thermal* process.

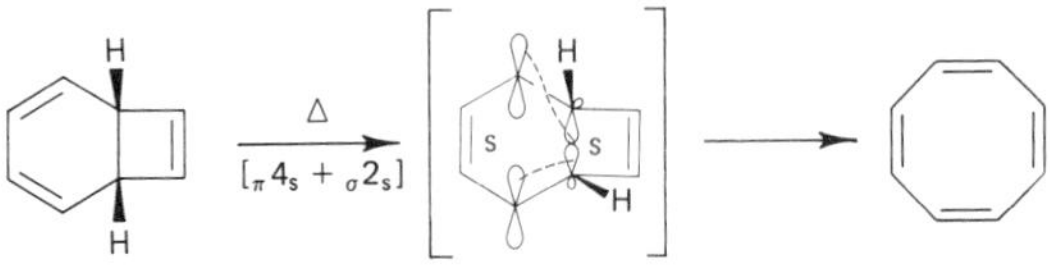

Δ $[_{\pi}4_s + _{\sigma}2_s]$

Transition state

The conversion of semibullvalene to cyclooctatetraene is symmetry forbidden to occur photochemically as a concerted process. The reaction is therefore a stepwise process, probably proceeding via a diradical intermediate.

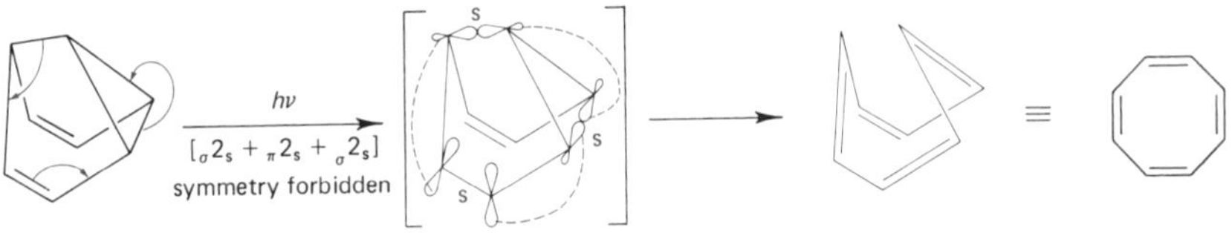

$h\nu$ $[_{\sigma}2_s + _{\pi}2_s + _{\sigma}2_s]$ symmetry forbidden

Transition state

PROBLEM V.4

Reference: M. Jones, Jr., *J. Amer. Chem. Soc.* **89**, 4236 (1967).
A mechanism for the photochemical conversion of bullvalene to (A) is shown below

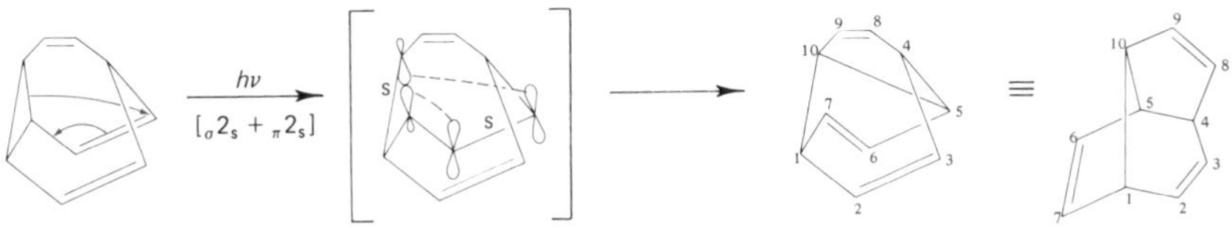

$h\nu$ $[_{\sigma}2_s + _{\pi}2_s]$

Transition state

This is an example of a vinylcyclopropane ⟶ cyclopentene rearrangement; the suprafacial [1,3] sigmatropic process is symmetry allowed to proceed photochemically in a concerted fashion.

Compound B can be formed from compound A via a $[_{\sigma}2_a + _{\pi}2_a]$ process (photochemically allowed). The formation of B from bullvalene probably proceeds sequentially via compound A rather than directly.

Bullvalene $\xrightarrow[[_\sigma 2_s + _\pi 2_s]]{h\nu}$ (A) $\xrightarrow[[_\sigma 2_a + _\pi 2_a]]{h\nu}$ [Transition state] $\longrightarrow$ (B)

The thermal conversion B $\longrightarrow$ C is expected to occur readily via a symmetry-allowed $[_\sigma 2_s + _\pi 2_s + _\sigma 2_s]$ pathway:

(B) $\xrightarrow{\Delta}$ [Transition state] $\longrightarrow$ $\equiv$ (C)

We would not expect the thermal conversion A $\longrightarrow$ C to be as facile as the corresponding conversion B $\longrightarrow$ C. In one possible route, A is first converted to B via the symmetry-*forbidden* thermal path, and then B $\longrightarrow$ C via the thermal pathway shown above.

ANSWERS TO PROBLEMS IN CHAPTER VI

PROBLEM VI.1

(a) *References*: W. L. Mock, *J. Amer. Chem. Soc.* **88**, 2857 (1966); S. D. McGregor and D. M. Lemal, *ibid.*, p. 2858.

$$\xrightarrow[{[{}_{\sigma}2_s + {}_{\pi}2_s + {}_{\sigma}2_s]}]{\Delta} SO_2 + H_3C\text{—CH=CH—CH=CH—}CH_3$$

This is a six-electron, linear, cheletropic* cycloreversion process. Disrotatory motion of the methyl groups accompanies expulsion of SO_2.

(b) *Reference*: J. J. Bloomfield, J. R. S. Irelan, and A. P. Marchand, *Tetrahedron Lett.* p. 5647 (1968).

The $[{}_{\sigma}2_s + {}_{\sigma}2_s]$ photochemically induced cycloreversion might occur.

$$\xrightarrow[?]{h\nu} + [\text{"}C_2O_2\text{"}] \longrightarrow + 2\,CO$$

This reaction is of theoretical interest as it affords a symmetry-allowed pathway to the hypothetical molecule, $O{=}C{=}C{=}O$. The question is thus raised: Is 2 CO produced directly via a stepwise (nonconcerted) photocycloreversion reaction, or might it be formed upon decomposition of the hypothetical intermediate $O{=}C{=}C{=}O$ which is itself initially formed via a concerted cycloreversion process? Thus far, attempts to isolate or trap $O{=}C{=}C{=}O$ have met with repeated failure.†

*A cheletropic process is one "... in which two σ-bonds which terminate at a single atom are made, or broken, in concert." For discussion, see R. B. Woodward and R. W. Hoffmann, "The Conservation of Orbital Symmetry," pp. 152–163. Academic Press, New York, 1970. See also Problem III.9.

† See J. Strating, B. Zwanenburg, A. Wagenaar, and A. C. Udding, *Tetrahedron Lett.* p. 125 (1969).

(c) *References*: G. O. Schenck and R. Steinmetz, *Tetrahedron Lett.* No. 21, p. 1 (1960); J. G. Atkinson, D. E. Ayer, G. Büchi, and E. W. Robb, *J. Amer. Chem. Soc.* **85**, 2257 (1963).

CN R C C R
$h\nu$
$[_{\pi}2_s + _{\pi}2_s]$
cycloaddition
CN R R H
$h\nu$
4-electron
electrocyclic process
disrotatory
CN R R
(A)

PROBLEM VI.2

References: R. F. Childs and V. Taguchi, *J. Chem. Soc.* (D), p. 695 (1970); R. K. Lustgarten, M. Brookhart, and S. Winstein, *J. Amer. Chem. Soc.* **89**, 6350 (1967).

(a)

H ⊕ H
(I)
$h\nu$
4-electron electrocyclic process
disrotatory
H-- H-- ⊕
(VI)

H-- H-- ⊕
[1,3] sigmatropic shift
$h\nu\ [_{\sigma}2_a + _{\pi}2_a]$
a H-- H-- a ⊕
Transition state
⊕ ≡ ⊕
(II)

The conversion VI ⟶ II can also be regarded as a two-electron process (1,2-shift).

H-- H-- ⊕
1,2-shift
Δ (<−60°C)
⊕ ≡ (II)

Hence, VI would be expected to be thermally labile as well (as is observed).

The thermal conversion II ⟶ I is symmetry forbidden to occur in a concerted fashion as a $[_{\sigma}2_a + _{\sigma}2_a + _{\pi}2_a]$ process.

⊕
(II)
Δ
+47° C
symmetry forbidden
⊕ a a a a
Transition state
⊕ ⟷ +
(I)

The relatively high temperature required for the rearrangement II ⟶ I (+47°C; cf. the rearrangement VI ⟶ II) is suggestive of the operation of a stepwise process.

(b)

hν
4-electron electrocyclic process
disrotatory

(IV)

hν
4-electron electrocyclic process
disrotatory

1,2-shift
Δ (−75° C)

H^+

$FSO_3^\ominus$

(V)

PROBLEM VI.3

Reference: J. Ciabattoni, J. E. Crowley, and A. S. Kende, *J. Amer. Chem. Soc.* **89**, 2778 (1967).

Compound A is formed via a Diels-Alder $[_\pi4_s + _\pi2_s]$ thermal reaction.

Δ
$[_\pi4_s + _\pi2_s]$

(A)

The conversion of A to B very likely occurs in two steps: first, a photochemically allowed suprafacial $[_\pi2_s + _\sigma2_s]$ [1,3] sigmatropic rearrangement to form B′, which in a second step readily aromatizes by two successive [1,3] hydrogen shifts.

hν
[1,3]

Transition state

(B′)

[1,3]

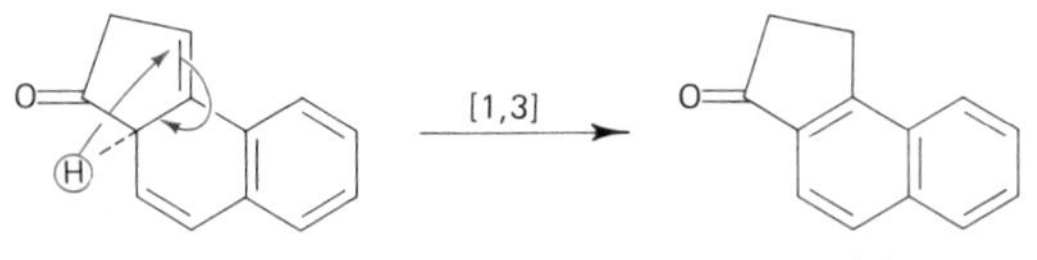

[1,3]

(B)

The aromatization process is photochemically allowed, or, equally likely, it may occur thermally by successive stepwise processes.

PROBLEM VI.4

Reference: G. L. Closs and P. E. Pfeffer, *J. Amer. Chem. Soc.* **90**, 2452 (1968).

The three possible products of this reaction are

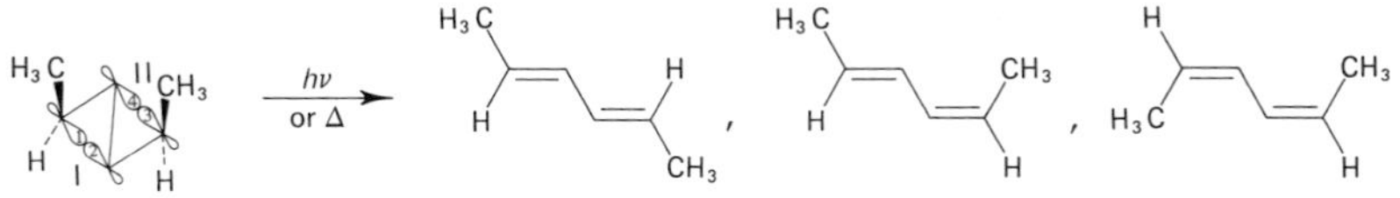

(A) —$h\nu$ or Δ→ *trans, trans* , *trans, cis* (same as *cis, trans*) , *cis, cis*

Consider the possibilities for reaction on A: inversion (i) or retention (r) at atoms 1, 2, 3, and 4; suprafacial (S) or antarafacial (A) at bonds I and II. With the aid of molecular models, the reader can show that there are basically only three different orbital motions which transform A into the *trans,trans-*, *trans,cis-*, and *cis,cis*-hexa-2,4-dienes. A representative of each type of transformation is indicated below as a guide:

Reactant	Classification	[$h\nu$ or Δ?]	Product
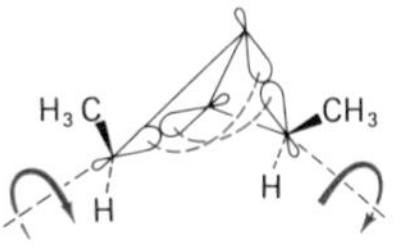	$[_{\sigma}2_s + _{\sigma}2_s]$	$h\nu$	 *cis, cis*
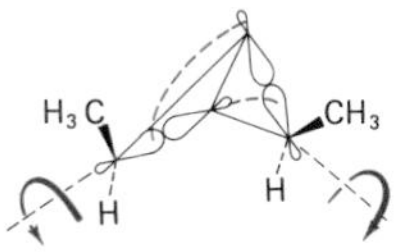	$[_{\sigma}2_a + _{\sigma}2_a]$	$h\nu$	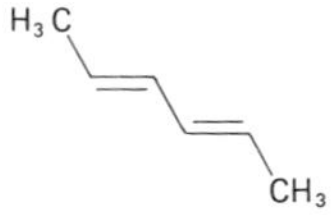 *trans, trans*
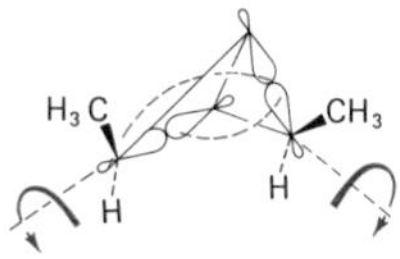	$[_{\sigma}2_a + _{\sigma}2_s]$	Δ	 *cis, trans* (*trans, cis*)

For the case *exo,exo*-2,4-dimethylbicyclobutane (A), the product is observed to have the *trans,cis* configuration, and, hence, is the product predicted to arise thermally (concerted $[_{\sigma}2_s + {_{\sigma}2_a}]$ process) on the basis of orbital symmetry considerations. The reader can verify that the *trans,trans*-olefin obtained from thermal rearrangement of *exo,endo*-2,4-dimethylbicylobutane is in accord with predictions based on orbital symmetry considerations.

PROBLEM VI.5

Reference: P. G. Gassman, *Accounts Chem. Res.* **3**, 26 (1970).

For a process in which ring opening and heterolysis of the nitrogen–chlorine bond are synchronous:

heterolysis of N – Cl bond proceeds with anchimeric assistance

2-electron process thermally disrotatory

(I) $R_1 = R_4 = H$
$R_2 = R_3 = CH_3$

(II) $R_1 = R_3 = H$
$R_2 = R_4 = CH_3$

$\xrightarrow{H_2O}$ + NH_4Cl

In the disrotatory ring opening, the sigma-bond which is undergoing rupture consequently buckles in such a manner as to anchimerically assist (backside displacement!) the solvolysis (heterolytic cleavage of the N–Cl bond). There is a larger relief of strain in the disrotatory ring opening of II (*cis*-methyl groups) than in the corresponding process in I (*trans*-methyl groups), and hence, $k_2 > k_1$.

The closely analogous carbocyclic system (cyclopropyl halide solvolysis) has been studied. Anchimeric assistance to solvolysis has been convincingly demonstrated in these systems (see Problem VII.2).

PROBLEM VI.6

Reference: W. von E. Doering and W. R. Roth, *Angew. Chem. Int. Ed. Engl.* **2**, 115 (1963).

This may be regarded as analogous to the butadiene–cyclobutene electrocyclic process. However, in the present example, one of the butadiene double bonds has been replaced by a cyclopropane ring. This then becomes a $[_{\sigma}2_s + {_{\pi}2_s}]$ process, and is hence thermally "forbidden" to proceed in a concerted fashion.

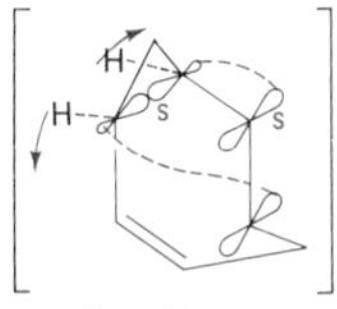

Transition state

A stepwise mechanism is therefore suggested, probably proceeding via a diradical intermediate.

PROBLEM VI.7

Reference: W. L. Dilling, R. D. Kroening, and J. C. Little, *J. Amer. Chem. Soc.* **92**, 928 (1970).

This probably proceeds in two steps. The first is symmetry "forbidden" to occur photochemically in a concerted fashion, and therefore proceeds in a stepwise fashion, probably via a diradical intermediate. The second step is photochemically allowed to proceed in a concerted fashion.

$h\nu$, sens.
$[_\pi 4_s + _\pi 2_s]$
$h\nu$ forbidden

(and *exo* isomer)

$h\nu$
$[_\pi 2_s + _\pi 2_s]$

Transition state

(and enantiomer)

PROBLEM VI.8

Reference: G. Schröder and J. F. M. Oth, *Tetrahedron Lett.* p. 4083 (1966).

This is a "double disrotation" which may be viewed in a number of alternative ways. One view might be to regard this as a photochemically allowed 16-electron $[_\pi 6_s + _\sigma 2_s + _\pi 6_s + _\sigma 2_s]$ process.

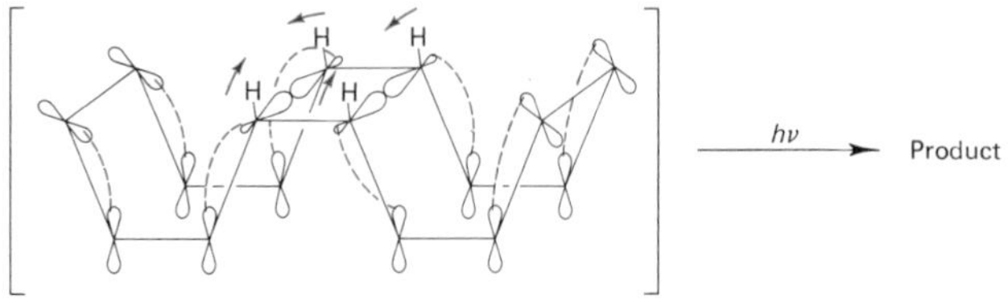

Transition state

Alternatively, we may regard the process purely in terms of the cyclobutane ring; this is a four-electron process, $[{}_{\sigma}2_s + {}_{\sigma}2_s]$, photochemically allowed.

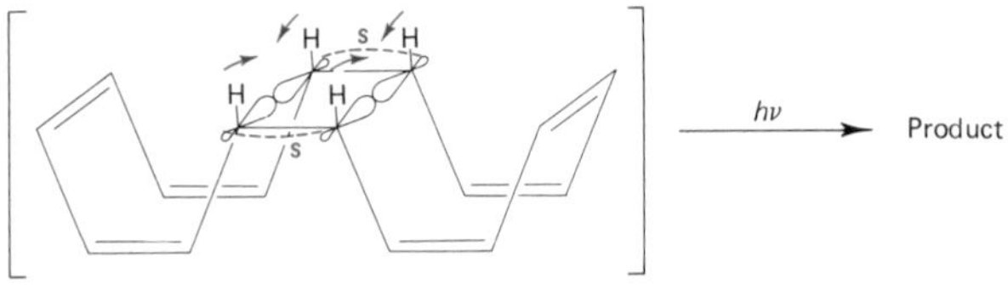

Transition state

A third possibility is that the reaction occurs in two concerted (photochemically allowed) eight-electron processes, $[{}_{\pi}6_s + {}_{\sigma}2_s]$.

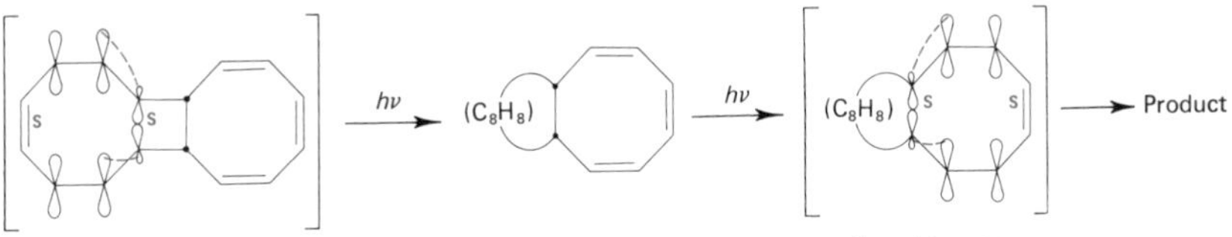

Transition state

Transition state

PROBLEM VI.9

Reference: J. A. Elix, M. V. Sargent, and F. Sondheimer, *Chem. Commun.* p. 509 (1966).

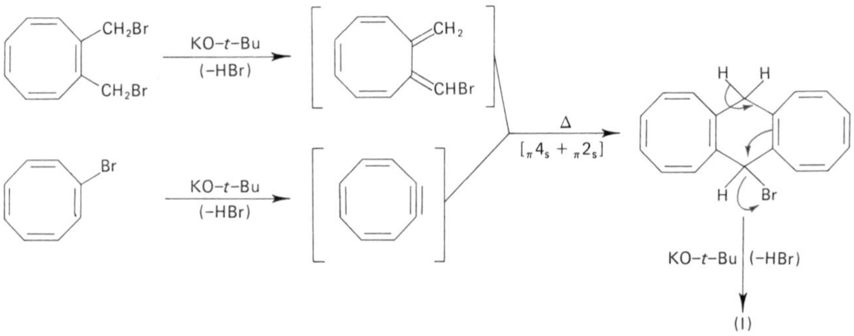

The periphery of I contains 18 pi-electrons, making it a *potentially* aromatic system. [Beware of applying the Hückel $(4n + 2)$ rule to polycyclic systems!] The question thus arises: Which structure, Ia or Ib, best represents compound I?

(Ia) (Ib)

NMR data for compound I are consistent only with structure Ib; apparently, the type of resonance interaction depicted in Ia is not important. (Note that the benzenoid aromaticity of the central six-membered ring is destroyed in Ia.) This result could be readily predicted using Dewar's PMO approach.*

$(a = 5^{-1/2})$

$$\begin{aligned} \delta E &= 2\beta\,(a_{or}\,b_{os} + a_{ot}\,b_{ou}) \\ &= 2\beta(5^{-1/2} \cdot 5^{-1/2} + 5^{-1/2} \cdot 5^{-1/2}) = \tfrac{4}{5}\beta \\ &= 0.80\beta \end{aligned}$$

for benzene: $(b = 3^{-1/2})$

$$\begin{aligned} \delta E &= 2\beta(1 \cdot 3^{-1/2} + 1 \cdot 3^{-1/2}) = 4 \cdot 3^{-1/2}\beta \\ &= 2.31\beta \end{aligned}$$

PROBLEM VI.10

Reference: M. Sakai, A. Diaz, and S. Winstein, *J. Amer. Chem. Soc.* **92**, 4452 (1970).

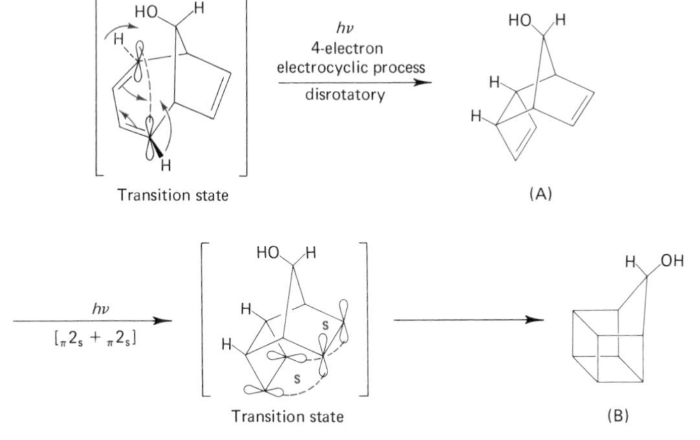

*See M. J. S. Dewar, "The Molecular Orbital Theory of Organic Chemistry," Chapter 6. McGraw-Hill, New York, 1969.

PROBLEM VI.11

Reference: W. B. Avila and R. A. Silva, *J. Chem. Soc.* (D), p. 94 (1970).

COOH C C COOH Δ $[_\pi 4_s + _\pi 2_s]$ COOH COOH , Δ $[_\pi 4_s + _\pi 2_s]$ COOH COOH $-H_2O$

O O O ≡ O O O $h\nu$ $[_\pi 2_s + _\pi 2_s]$ O O O

(A) (B)

PROBLEM VI.12

Reference: C. A. Cupas, W. Schumann, and W. E. Heyd, *J. Amer. Chem. Soc.* **92**, 3237 (1970).

The reaction can take place via a series of [1,5] sigmatropic hydrogen shifts followed by a Claisen rearrangement and an intramolecular Diels-Alder reaction.

H O Δ [1,5] O H H [1,5] H O H [1,5] O H H

[3,3] Claisen rearrangement O H H ≡ O H H

(C)

Δ $[_\pi 4_s + _\pi 2_s]$ 10 9 8 6 7 5 4 3 2 1 O ≡ 8 6 O 5 7 4 10 3 9 1 2

(A)

The formation of B from intermediate C can occur in the following way:

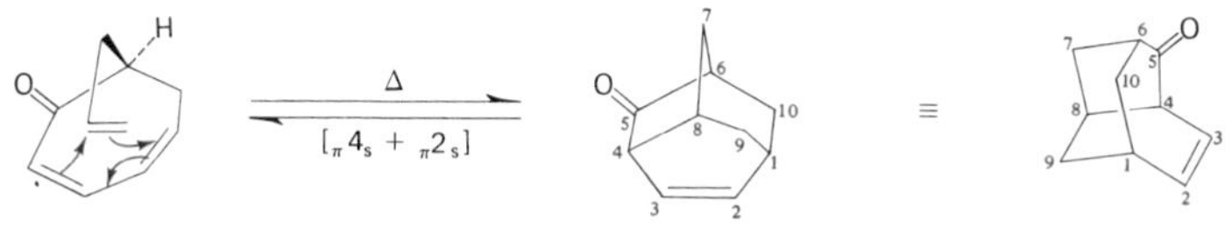

(C) $\xrightarrow[[_{\pi}4_s + _{\pi}2_s]]{\Delta}$ ≡ (B)

Interestingly, there are three other products which might possibly have been formed in this reaction:

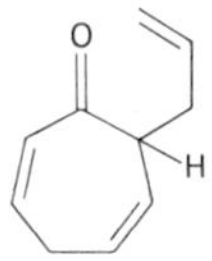

(I) 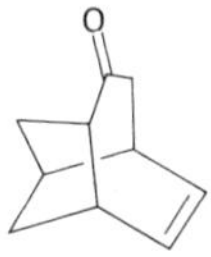 (II) 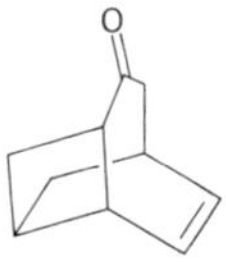 (III)

These, however, are not observed among the final products of the reaction. Can you suggest reasons why A and B are formed to the exclusion of I, II, and III?

PROBLEM VI.13

References: R. Huisgen and W. E. Konz, *J. Amer. Chem. Soc.* **92**, 4102 (1970); W. E. Konz, W. Hechtl, and R. Huisgen, *ibid.*, p. 4104; R. Huisgen, W. E. Konz, and G. E. Gream, *ibid.*, p. 4105.

Observation (a) suggests that there is an ionization step whose rate constant figures in the overall kinetic equation; the ionization proceeds more rapidly in

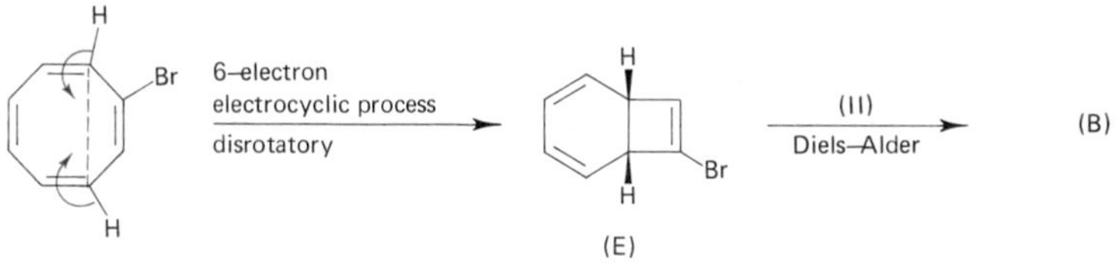

6–electron electrocyclic process disrotatory → (E) → (II) Diels–Alder → (B)

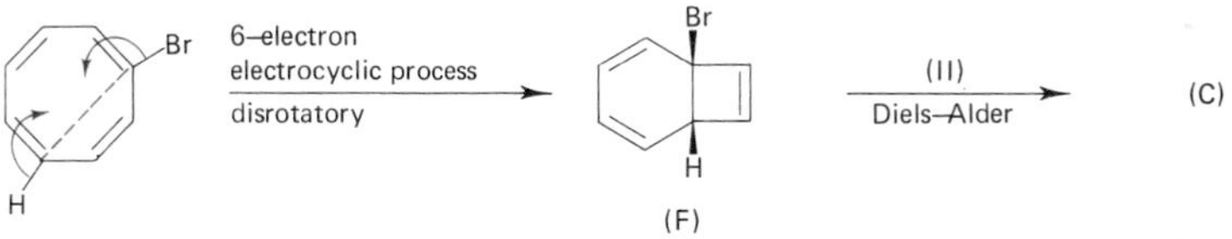

6–electron electrocyclic process disrotatory → (F) → (II) Diels–Alder → (C)

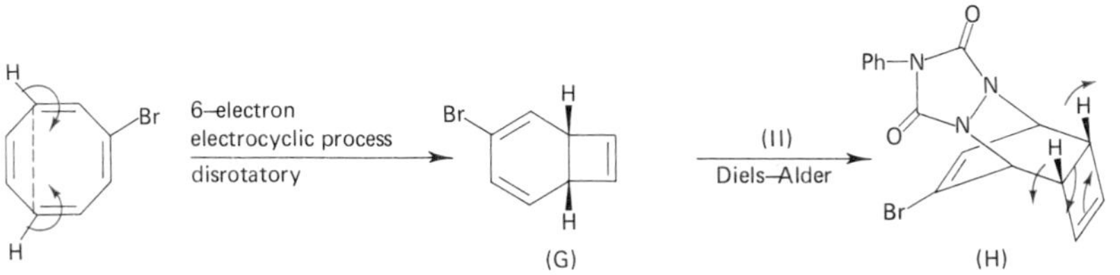

6–electron electrocyclic process disrotatory → (G) → (II) Diels–Alder → (H)

(H) 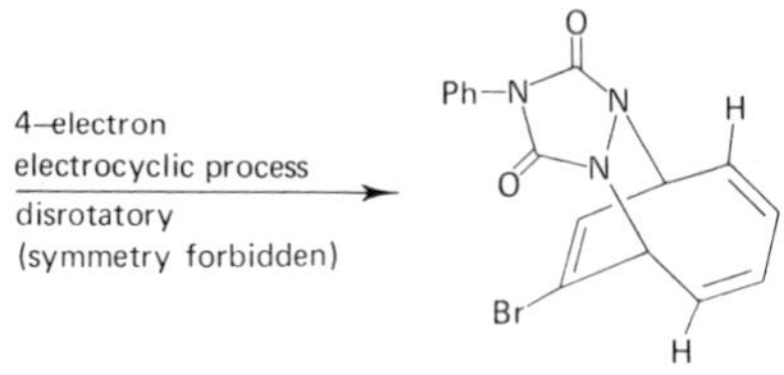 —4-electron electrocyclic process, disrotatory (symmetry forbidden)→ (D)

the more highly polar ("better-solvating") solvent. Observation (b) suggests the intermediacy of structures E, F, and G, which are trapped by the triazolinedione dienophile to afford the observed products, B, C, and D, respectively.

The rearrangement III ⟶ IV [observation (c)] can be rationalized in terms of the formation of an intermediate analogous to F.

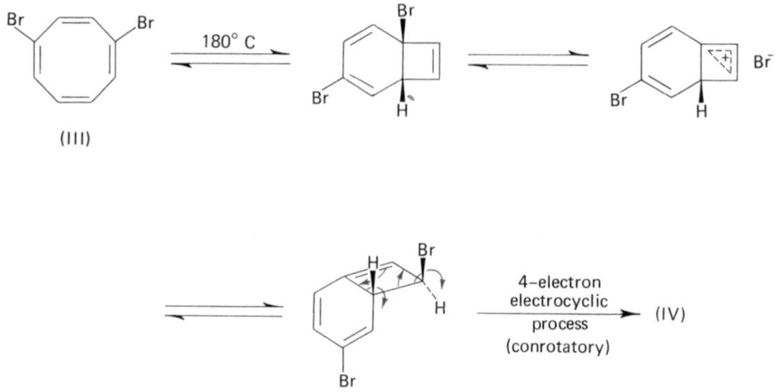

(III) —180° C⇌ ⇌ Br^-

⇌ —4-electron electrocyclic process (conrotatory)→ (IV)

The [1,3] suprafacial bromine shift is, of course, a symmetry-"forbidden" thermal process; the suggested ionization (stepwise) process is consistent with the observed solvent effect [observation (a)].

Finally, we may picture the rearrangement I ⟶ A as follows:

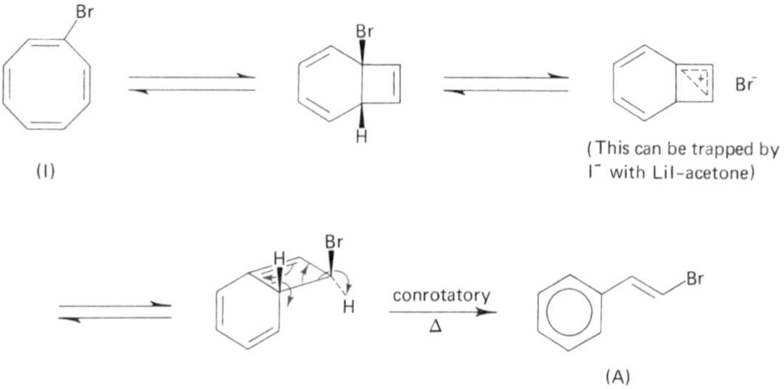

(I) ⇌ ⇌ Br^-

(This can be trapped by I^- with LiI-acetone)

⇌ —conrotatory, Δ→ (A)

PROBLEM VI.14

References: K. Alder, F. H. Flock, and P. Janssen, *Chem. Ber.* **89**, 2689 (1956); T. J. Katz, M. Rosenberger, and R. K. O'Hara, *J. Amer. Chem. Soc.* **86**, 249 (1964).

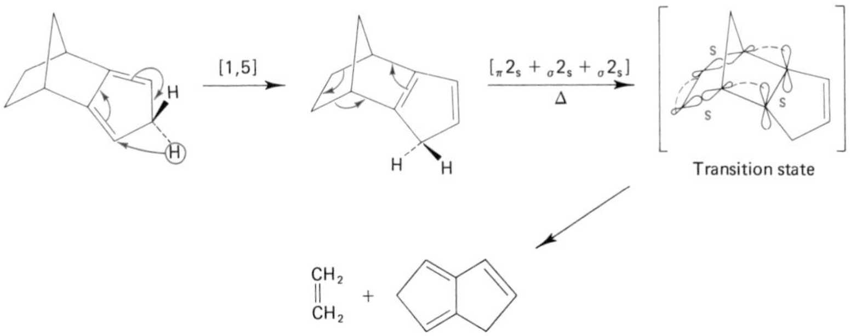

[1,5]

$[_{\pi}2_s + _{\sigma}2_s + _{\sigma}2_s]$ Δ

Transition state

$CH_2{=}CH_2$ +

PROBLEM VI.15

Reference: T. Mukai and K. Kurabayashi, *J. Amer. Chem. Soc.* **92**, 4493 (1970). This is an eight-electron, linear, thermal, cheletropic cycloreversion; extrusion of CO is, accordingly, a symmetry-allowed photochemical process. A reasonable mechanism for this process is indicated below:

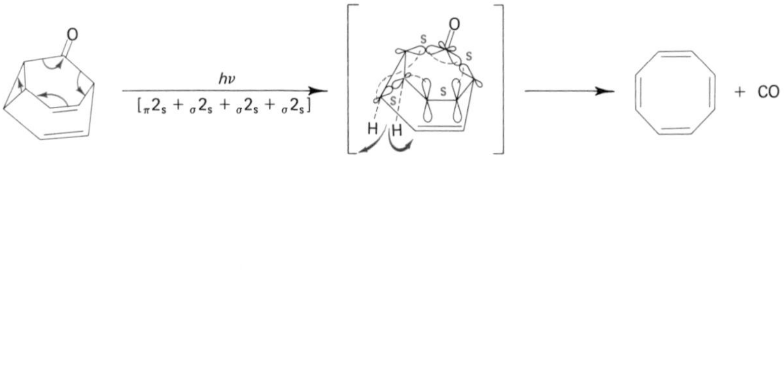

$h\nu$ $[_{\pi}2_s + _{\sigma}2_s + _{\sigma}2_s + _{\sigma}2_s]$

+ CO

ANSWERS TO PROBLEMS IN CHAPTER VII

PROBLEM VII.1

(a) *Reference*: I. Fleming and E. Wildsmith, *J. Chem. Soc.* (D), p. 223 (1970).

(1) Δ disrotatory

Δ $MeO_2C—C≡C—CO_2Me$ (Diels-Alder)

CO_2Me CO_2Me (A)

D_2 Pd/C

D D CO_2Me CO_2Me (B)

(B) 200°C $[_\sigma 2_s + _\sigma 2_s + _\pi 2_s]$ cycloreversion CO_2Me CO_2Me + (C) Δ conrotatory (D)

(1) Δ, maleic anhydride (2) H_2O

HO_2C HO_2C H D D H (E)

(E) $Pb(OAc)_4$ pyridine H D D H (F) Δ $[_\sigma 2_s + _\sigma 2_s + _\pi 2_s]$ cycloreversion D + HD

(2) O D D Δ −CO D D (G) Δ $[_\sigma 2_s + _\sigma 2_s]$ cycloreversion (thermally "forbidden") D(H) + HD + D_2 + H_2

(b) The thermal dehydrogenation of F is a symmetry-allowed $[_{\sigma}2_s + _{\sigma}2_s + _{\pi}2_s]$ cycloreversion, which occurs cleanly to give HD and benzene-d_1. The thermal dehydrogenation of G is a symmetry-"forbidden" $[_{\sigma}2_s + _{\sigma}2_s]$ process; the fact that it occurs nonstereospecifically suggests that it may occur via a stepwise (radical) process.

PROBLEM VII.2

(a) *Reference*: C. B. Reese and A. Shaw, *J. Amer. Chem. Soc.* **92**, 2566 (1970).

In the transition state of the solvolysis reaction, the *p* orbitals which develop during the cyclopropane carbon–carbon bond cleavage process anchimerically assist the solvolytic cleavage of the carbon–bromine bond *stereospecifically* via an intramolecular backside attack (S_N2 displacement) on the carbon atom bearing the leaving group (Br). Note that such stereospecificity requires that cyclopropane C—C bond cleavage and C—Br bond heterolysis must be concurrent; were the overall reaction to occur in a stepwise fashion (solvolysis followed by ring opening), reactions (1) and (2) would necessarily proceed via the same intermediate carbonium ion, and hence would afford the same product or mixture of products (as is clearly *not* the case).

The ring-opening process is a two-electron electrocyclic process. Two disrotatory paths are therefore symmetry allowed to occur thermally; these are illustrated below:

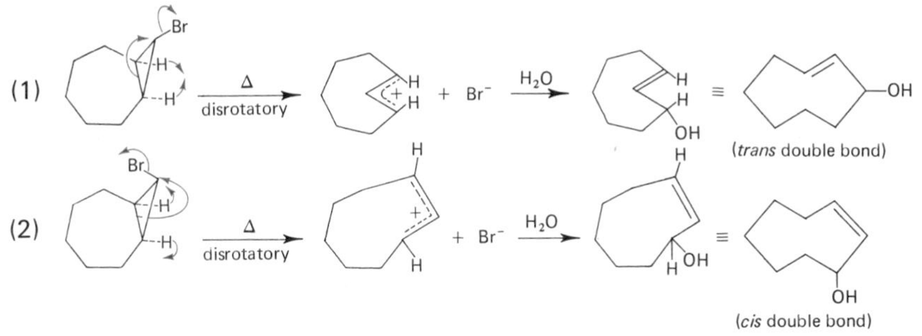

Of these two processes, it can be shown that the one leading to formation of the *trans* double bond is sterically preferred.* Hence, reaction (1) proceeds at a more rapid rate than does reaction (2).

(b) *Reference*: P. S. Skell and S. R. Sandler, *J. Amer. Chem. Soc.* **80**, 2024 (1958).

*C. H. DePuy, *Accounts Chem. Res.* **1**, 33 (1968).

For each compound, A and B, we must *a priori* consider the two possible modes of symmetry-allowed thermal disrotatory ring opening.

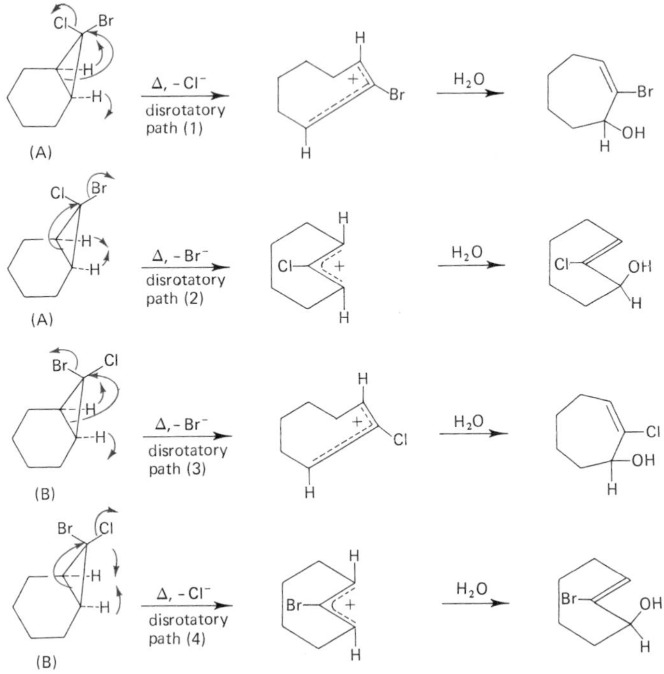

(A) Δ, - Cl^- disrotatory path (1) → H_2O →

(A) Δ, - Br^- disrotatory path (2) → H_2O →

(B) Δ, - Br^- disrotatory path (3) → H_2O →

(B) Δ, - Cl^- disrotatory path (4) → H_2O →

We do not expect paths (2) and (4) to be operative; it is not energetically feasible to produce a *trans*-cycloheptene! Hence, we distinguish between A and B solvolytically by examination of the products formed by paths (1) and (3). Compound A affords the bromine-containing cycloheptanol whereas solvolysis of compound B affords the corresponding chlorine-containing products.

(c)

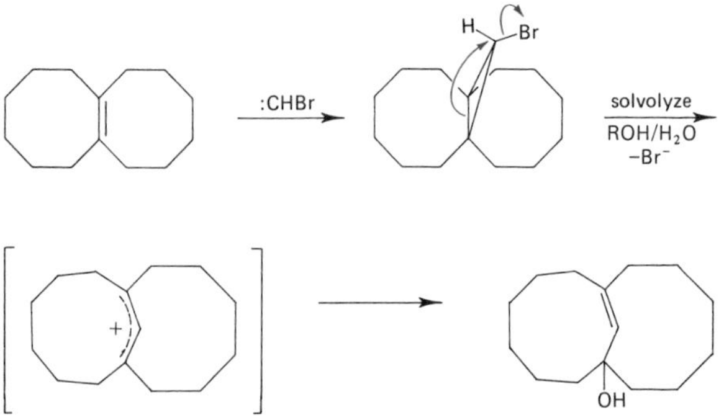

:CHBr → solvolyze ROH/H_2O $-Br^-$ →

→ (C)

PROBLEM VII.3

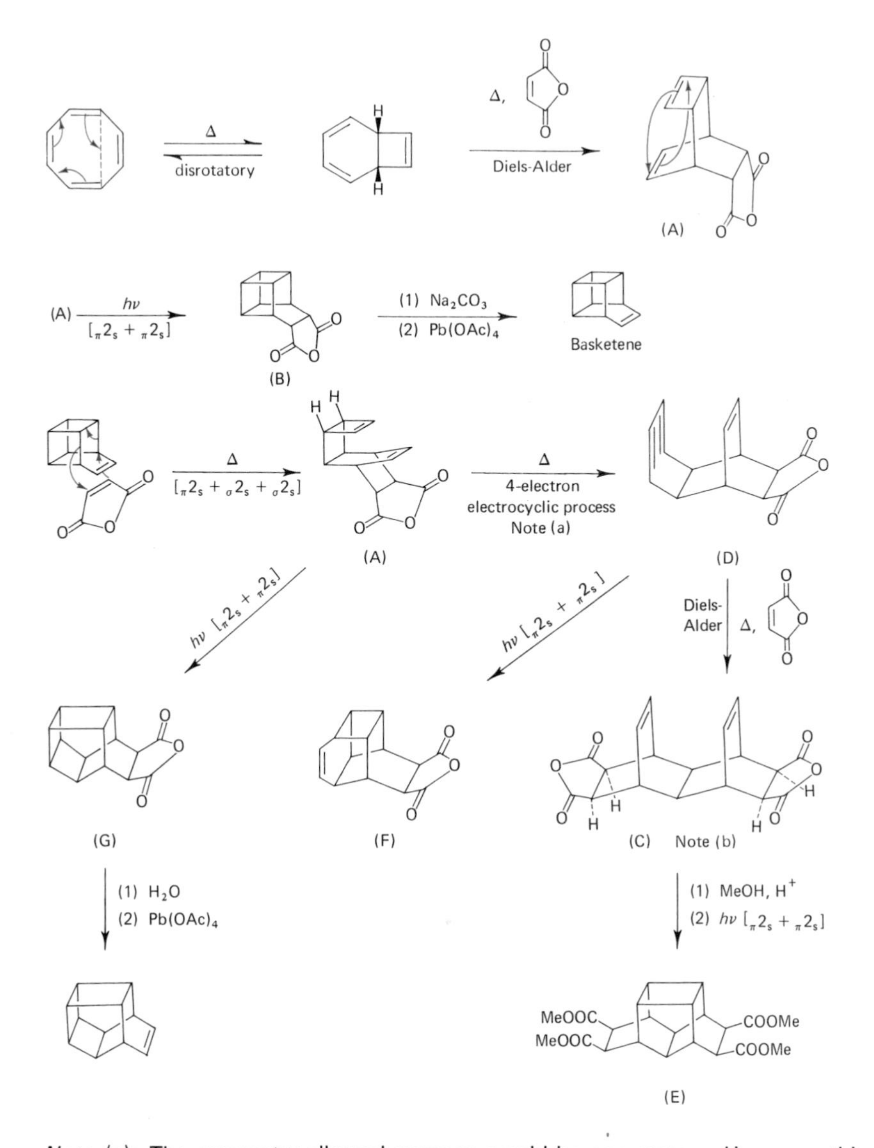

Note (*a*): The symmetry-allowed process would be conrotatory. However, this would require formation of a *trans*-cyclohexene. Hence, this reaction most likely occurs via a nonconcerted (stepwise) process.

Note (*b*): The *exo, exo* configuration of C is suggested by its facile photochemical transformation to E. The *exo, exo* conformation of the anhydride groups in C is consistent with chemical and spectroscopic information given by LeGoff [*J. Amer. Chem. Soc.* **91**, 5665 (1969)].

PROBLEM VII.4

Reference: See E. Vogel, *Chem. Soc. Spec. Publ.* No. 21, 113–147 (1967).

(a) This is a six-electron electrocyclic process, a symmetry-allowed thermally disrotatory process.

Δ
disrotatory

(III) (I)

The disrotatory process occurs readily as the geometrical requirements of the ring system offer no substantial hindrance. Note, however, that the corresponding conrotatory process is *not* possible in this system!

(b) *Reference*: E. Vogel, W. Grimme, and S. Korte, *Tetrahedron Lett.* p. 3625 (1965).

Δ

Δ, MeOOC—C≡C—COOMe
Diels-Alder

MeOOC
MeOOC

(I) (III) (A)

(A) Δ $[_{\pi}2_s + _{\sigma}2_s + _{\sigma}2_s]$ cycloreversion

COOMe
COOMe
+

(B)

It is apparent that compound B is an extremely highly strained molecule. One might well ask, "Why does compound B exist at all?" The answer is that further reactions of B appear to afford products which are thermodynamically less favorable than is B itself. Some unsavory alternatives are illustrated:

(1)

?

+ $:CH_2$

This is formally a "(2 + 0)" cycloreversion; the reaction as shown is probably endothermic.

(2) $\xrightarrow{?}$ 2 HC≡CH + $\left(\xrightarrow{?} C{\equiv}C + :CH_2\right)$

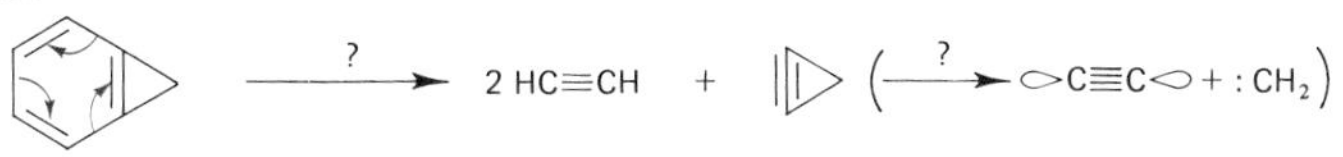

(3) 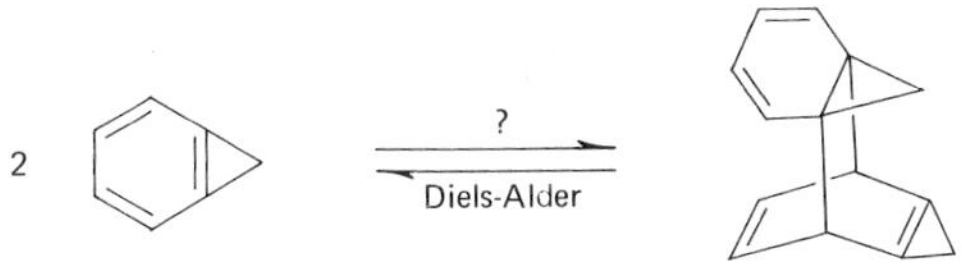 2 $\underset{\text{Diels-Alder}}{\overset{?}{\rightleftharpoons}}$

This is a symmetry-allowed thermal cycloaddition reaction. However, it is thermodynamically unfavorable as benzenoid resonance in *two* phenyl rings is lost in proceeding from reactants to product. Although probably less strained than reactants, the adduct is nevertheless a highly strained molecule. We therefore expect this to be an endothermic process with the equilibrium lying far to the left.

(c) *Reference*: E. Vogel, F. Weyres, H. Lepper, and V. Rautenstrauch, *Angew. Chem.* **78**, 754 (1966).

The observed products of the attempted MnO_2 oxidation of V were naphthalene and carbon monoxide. These could arise via cheletropic fragmentation of the intermediate ketone, IV.

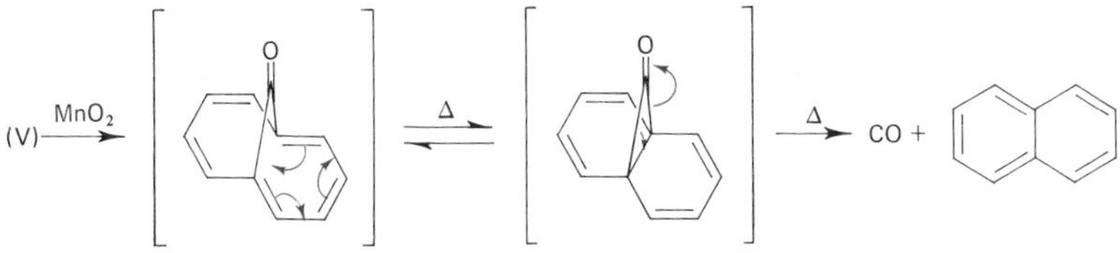

(V) $\xrightarrow{MnO_2}$ [ ] $\underset{}{\overset{\Delta}{\rightleftharpoons}}$ [ ] $\xrightarrow{\Delta}$ CO +

Not isolated Not isolated

The indicated four-electron process is symmetry allowed to occur thermally via a concerted nonlinear cheletropic process. The thermodynamic stability of the products, naphthalene and CO, provides ample driving force for the concerted decarbonylation process.

(d) *Reference*: E. Vogel, R. Feldmann, and H. Düwel, *Tetrahedron Lett.* p. 1941 (1970).

CHO CHO + $Ph_3\overset{+}{P}-\overset{-}{C}H$ CH_2 $Ph_3\overset{+}{P}-\overset{-}{C}H$ $\xrightarrow[60°C]{DMF}$ (C) $\xrightarrow[CH_2Cl_2]{Ph_3C^+BF_4^-}$ (VI) BF_4^-

(I) $\xrightarrow[CuCl]{CH_2N_2}$ [(D)] $\xrightarrow{\Delta}$ (E) $\xrightarrow{Ph_3C^+BF_4^-}$ (VI)

The interconversion D ⇌ E is a 10-electron electrocyclic process (thermally disrotatory).

PROBLEM VII.5

(a) *Reference*: P. Dowd, *J. Amer. Chem. Soc.* **92**, 1066 (1970).

$\xrightarrow{\Delta,\ SO_2}$ 6-electron cheletropic process $\xrightarrow{CrO_3}$ $\xrightarrow[-SO_2]{\Delta}$ (A) $\xrightarrow[-CO]{h\nu}$ (B)

(A)

(B)
Tetramethyleneethane

The photochemical conversion (A) ⟶ (B) affords tetramethyleneethane, which is predicted by simple molecular orbital theory to be a ground-state triplet (diradical).

(B)

E: $\alpha-2\beta$, $\alpha-\beta$, α, $\alpha+\beta$, $\alpha+2\beta$

(b) *Reference*: J. J. Gajewski and C.-N. Shih, *J. Amer. Chem. Soc.* **89**, 4532 (1967).

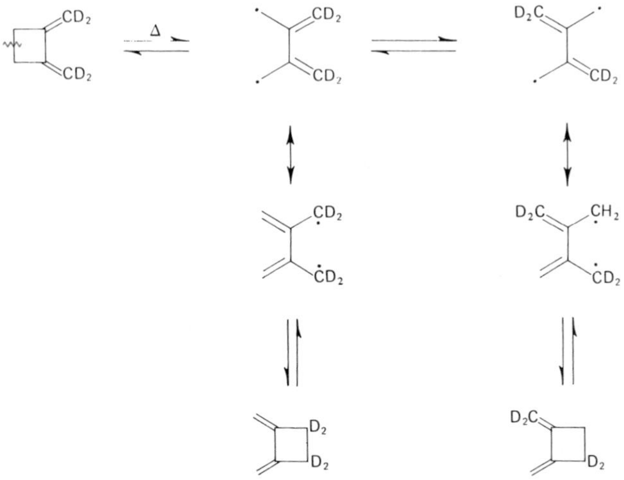

CD_2 CD_2 Δ CD_2 CD_2 D_2C CD_2

CD_2 CD_2 D_2C CH_2 CD_2

D_2 D_2 D_2C D_2

(c) This result rules out a mechanism whereby the reactants suffer symmetry-"forbidden" thermal $[_{\sigma}2_s + _{\sigma}2_s]$ cycloreversion to allene fragments followed by redimerization of the fragments to form the observed products:

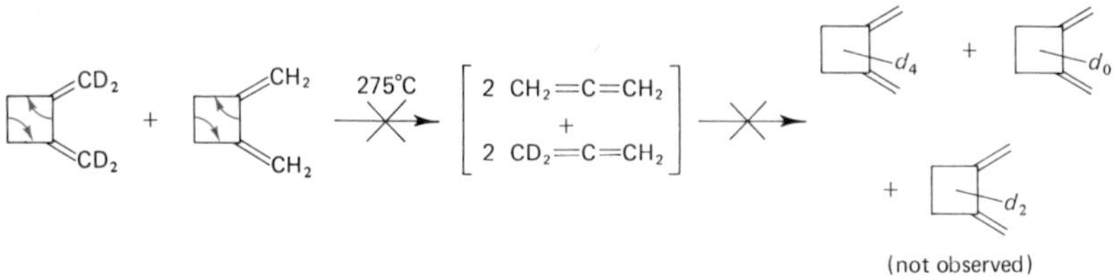

CD_2 CD_2 + CH_2 CH_2 275°C

$$\left[\begin{array}{c} 2\ CH_2{=}C{=}CH_2 \\ + \\ 2\ CD_2{=}C{=}CH_2 \end{array}\right]$$

d_4 + d_0 + d_2

(not observed)

PROBLEM VII.6

Reference: J. A. Elix, M. V. Sargent, and F. Sondheimer, *Chem. Commun.* pp. 508 and 509 (1966).

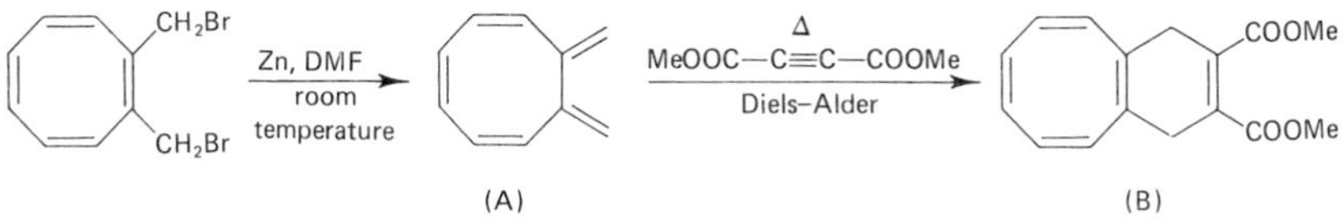

CH_2Br CH_2Br Zn, DMF room temperature

Δ MeOOC—C≡C—COOMe Diels–Alder

COOMe COOMe

(A) (B)

(B) —[DDQ (2,3-dichloro-5,6-dicyano-p-benzoquinone), Δ; $-H_2$]→ (C)

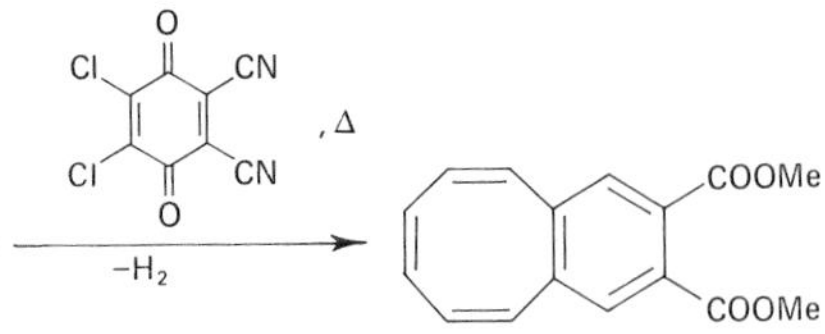

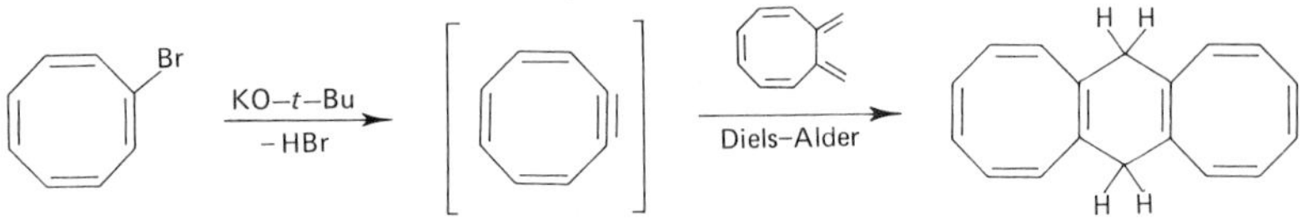

Br-cyclooctatetraene —[KO–t–Bu; $-$HBr]→ [(D)] —[Diels–Alder]→ (E)

(E) —[$-H_2$]→ (F)

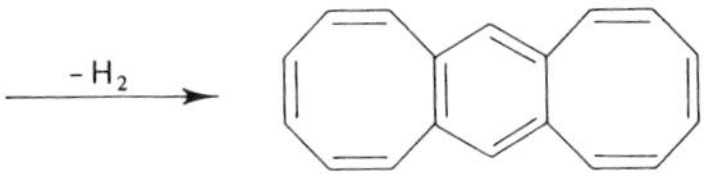

PROBLEM VII.7

Reference: G. Büchi and J. E. Powell, Jr., *J. Amer. Chem. Soc.* **92**, 3126 (1970).

(a) This is a [3,3] sigmatropic rearrangement (Claisen rearrangement).

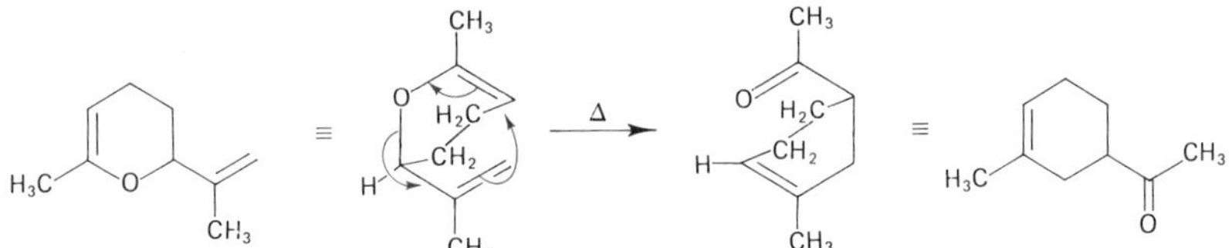

(C) ≡ [transition-state drawing] —Δ→ ≡ (A)

Note that the fact that A is produced free from contamination by B argues against dissociation of the starting material (C) to isoprene and methyl vinyl ketone, followed by recombination via Diels-Alder reaction as an alternative mechanism.

(b) The two possible transition states are "chair-like" and "boat-like."

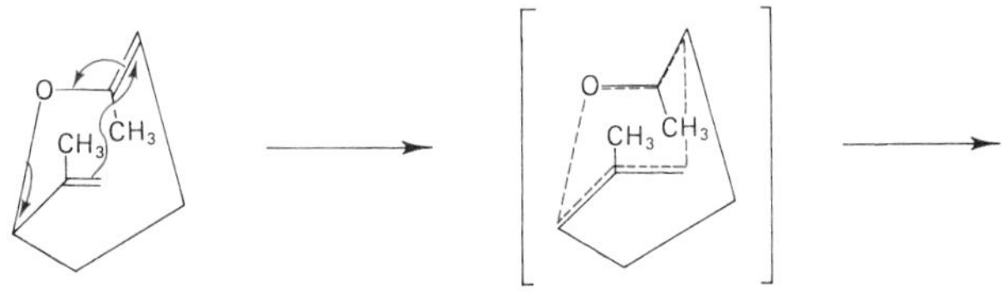

→ [ ] → (A)

"Chair-like"
transition state

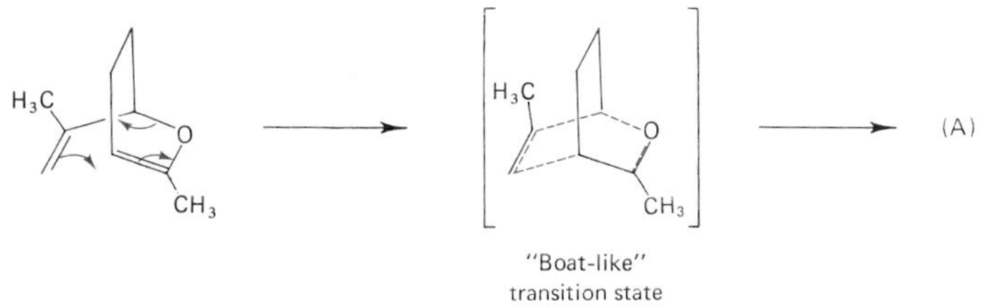

"Boat-like" transition state

Inspection of molecular models clearly reveals that the rearrangement must proceed through a "boat-like" transition state. This is a particularly interesting result in view of the classic results of Doering and Roth* who showed some time ago that ordinarily the "chair-like" transition state is preferred in *unstrained* [3,3] sigmatropic rearrangements.

PROBLEM VII.8

Reference: K. B. Wiberg, V. Z. Williams, Jr., and L. E. Friedrich, *J. Amer. Chem. Soc.* **92**, 564 (1970).

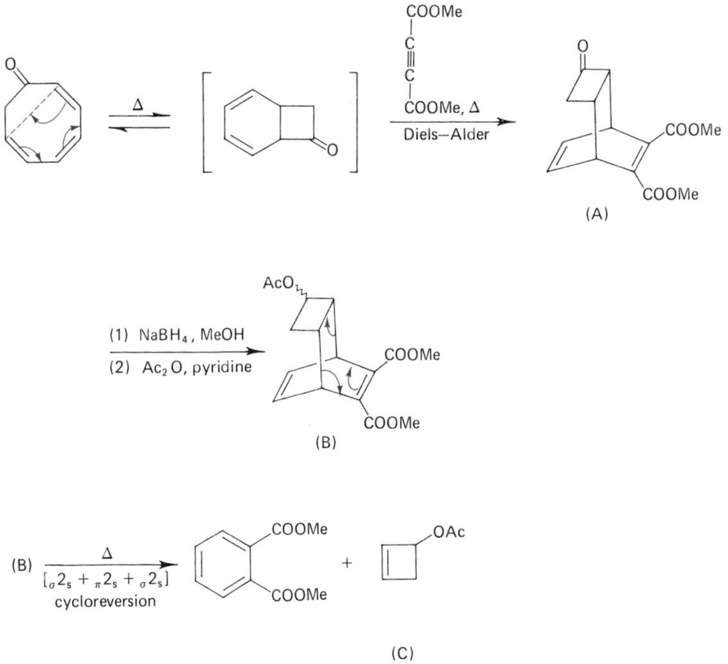

(1) $NaBH_4$, MeOH
(2) Ac_2O, pyridine

(B) $\xrightarrow[\left[{}_{\sigma}2_s + {}_{\pi}2_s + {}_{\sigma}2_s\right] \text{ cycloreversion}]{\Delta}$

*W. von E. Doering and W. R. Roth, *Tetrahedron* **18**, 67 (1962).

trans-1-Acetoxybutadiene is formed thermally from compound C via a four-electron electrocyclic process (conrotatory ring opening).

OAc H H H Δ conrotatory OAc H

ANSWERS TO PROBLEMS IN CHAPTER VIII

PROBLEM VIII.1

K. B. Wiberg, G. J. Burgmaier, and P. Warner, *J. Amer. Chem. Soc.* **93**, 246 (1971).

PROBLEM VIII.2

A. Krantz. *Nat. Meeting Amer. Chem. Soc. 161st, Los Angeles, California, March 29–April 2, 1971, Abstr. Pap.* paper no. ORGN 135.

PROBLEM VIII.3

W. H. Dolbier, Jr., and S.-H. Dai, *J. Chem. Soc.* (D), p. 166 (1971).

PROBLEM VIII.4

R. C. Cookson, J. Hudec, and M. Sharma, *J. Chem. Soc.* (D), pp. 107, 108 (1971).

PROBLEM VIII.5

N. S. Bhacca, L. J. Luskus, and K. N. Houk, *J. Chem. Soc.* (D), p. 109 (1971).

PROBLEM VIII.6

J. A. Berson, R. R. Boettcher, and J. J. Vollmer, *J. Amer. Chem. Soc.* **93**, 1540 (1971).

PROBLEM VIII.7

L. A. Paquette and J. C. Stowell, *J. Amer. Chem. Soc.* **93**, 2459 (1971).

PROBLEM VIII.8

B. Fuchs, *J. Amer. Chem. Soc.* **93**, 2544 (1971).

PROBLEM VIII.9

R. C. Cookson and J. E. Kemp, *J. Chem. Soc.* (D), p. 385 (1971).

PROBLEM VIII.10

(a) J. A. Berson and N. M. Hasty, Jr., *J. Amer. Chem. Soc.* **93**, 1549 (1971).
(b) P. Vogel, M. Saunders, N. M. Hasty, Jr., and J. A. Berson, *J. Amer. Chem. Soc.* **93**, 1551 (1971).

PROBLEM VIII.11

(a)

(1) H. H. Westberg and H. Ona, *J. Chem. Soc.* (D), p. 248 (1971); L. A. Paquette, R. S. Beckley, and T. McCreadie, *Tetrahedron Lett.* p. 775 (1971).
(2) L. Cassar, P. E. Eaton, and J. Halpern, *J. Amer. Chem. Soc.* **92**, 3515, 6366 (1970).
(3) J. Wristers, L. Brenner, and R. Pettit, *J. Amer. Chem. Soc.* **92**, 7499 (1970).
(4) L. A. Paquette, G. R. Allen, Jr., and R. P. Henzel, *J. Amer. Chem. Soc.* **92**, 7002 (1970).
(5) K. L. Kaiser, R. F. Childs, and P. M. Maitlis, *J. Amer. Chem. Soc.* **93**, 1270 (1971).
(6) T. J. Katz and S. Cerefice, *J. Amer. Chem. Soc.* **91**, 2405, 6519 (1969); **93**, 1049 (1971).

(b)

(1) *Uncatalyzed Rearrangement*: G. L. Closs and P. E. Pfeffer, *J. Amer. Chem. Soc.* **90**, 2452 (1968); *Ag(I) Catalyzed Rearrangement*: M. Sakai, H. Yamaguchi, H. H. Westberg, and S. Masamune, *J. Amer. Chem. Soc.* **93**, 1043 (1971).
(2) L. A. Paquette, R. P. Henzel, and S. E. Wilson, *J. Amer. Chem. Soc.*, **93**, 2336 (1971).
(3) L. Skattebøl, *Tetrahedron Lett.* p. 2361 (1970); W. R. Moore, K. G. Taylor, P. Müller, S. S. Hall, and Z. L. F. Gaibel, *Tetrahedron Lett.* p. 2365 (1970); P. G. Gassman and F. J. Williams, *J. Amer. Chem. Soc.* **92**, 7631 (1970).

(4) P. G. Gassman, T. J. Atkins, and F. J. Williams, *J. Amer. Chem. Soc.* **93**, 1812 (1971).

(c)

(1) W. G. Dauben, C. H. Schallhorn, and D. L. Whalen, *J. Amer. Chem. Soc.* **93**, 1446 (1971).

(2) L. A. Paquette, *J. Amer. Chem. Soc.* **92**, 5765 (1970).

(3) P. E. Eaton and S. A. Cerefice, *J. Chem. Soc.* (D), p. 1494 (1971).

The following references deal with proposed mechanisms for transition metal-catalyzed pericyclic reactions. The reader is referred to these papers for a comprehensive discussion of mechanism.

F. D. Mango, *Tetrahedron Lett.* p. 505 (1971).

G. S. Lewandos and R. Pettit, *Tetrahedron Lett.* p. 789 (1971).

F. D. Mango and J. H. Schachtschneider, *J. Amer. Chem. Soc.* **93**, 1123 (1971).

L. A. Paquette, *Accounts Chem. Res.* **4**, 280 (1971).

AUTHOR INDEX

After the author's name appears the problem number followed by the page number in parentheses on which that problem appears. References to authors whose names appear in footnotes and/or in suggested reading lists appear in this index as italicized numbers in parentheses.